Shouyang Wang · Yusen Xia

Portfolio Selection and Asset Pricing

332.6
W24p

Authors

Shouyang Wang
Academy of Mathematics and Systems Sciences
Chinese Academy of Sciences
Beijing 100080, China

Xia Yusen
McCombs School of Business
University of Texas at Austin
TX 78712 Austin, USA

Cataloging-in-Publication data applied for

Die Deutsche Bibliothek - CIP-Einheitsaufnahme

Wang, Shouyang:
Portfolio selection and asset pricing / Shouyang Wang ; Yusen Xia. - Berlin ; Heidelberg ; New York ; Barcelona ; Hong Kong ; London ; Mailand ; Paris ; Tokyo : Springer, 2002
(Lecture notes in economics and mathematical systems ; Vol. 514)
ISBN 3-540-42915-8

ISSN 0075-8450
ISBN 3-540-42915-8 Springer-Verlag Berlin Heidelberg New York

This work is subject to copyright. All rights are reserved, whether the whole or part of the material is concerned, specifically the rights of translation, reprinting, re-use of illustrations, recitation, broadcasting, reproduction on microfilms or in any other way, and storage in data banks. Duplication of this publication or parts thereof is permitted only under the provisions of the German Copyright Law of September 9, 1965, in its current version, and permission for use must always be obtained from Springer-Verlag. Violations are liable for prosecution under the German Copyright Law.

Springer-Verlag Berlin Heidelberg New York
a member of BertelsmannSpringer Science+Business Media GmbH

http://www.springer.de

© Springer-Verlag Berlin Heidelberg 2002
Printed in Germany

The use of general descriptive names, registered names, trademarks, etc. in this publication does not imply, even in the absence of a specific statement, that such names are exempt from the relevant protective laws and regulations and therefore free for general use.

Typesetting: Camera ready by author
Cover design: *design & production*, Heidelberg

Printed on acid-free paper SPIN: 10858324 55/3142/du 5 4 3 2 1 0

Preface

In our daily life, almost every family owns a portfolio of assets. This portfolio could contain real assets such as a car, or a house, as well as financial assets such as stocks, bonds or futures. Portfolio theory deals with how to form a satisfied portfolio among an enormous number of assets. Originally proposed by H. Markowtiz in 1952, the mean-variance methodology for portfolio optimization has been central to the research activities in this area and has served as a basis for the development of modern financial theory during the past four decades. Follow-on work with this approach has born much fruit for this field of study. Among all those research fruits, the most important is the capital asset pricing model (CAPM) proposed by Sharpe in 1964. This model greatly simplifies the input for portfolio selection and makes the mean-variance methodology into a practical application. Consequently, lots of models were proposed to price the capital assets.

In this book, some of the most important progresses in portfolio theory are surveyed and a few new models for portfolio selection are presented. Models for asset pricing are illustrated and the empirical tests of CAPM for China's stock markets are made.

The first chapter surveys ideas and principles of modeling the investment decision process of economic agents. It starts with the Markowitz criteria of formulating return and risk as mean and variance and then looks into other related criteria which are based on probability assumptions on future prices of securities. This chapter also presents methodologies which, instead of assuming probability distributions, rely on the best solution for the worst case scenario or in the average.

University Libraries
Carnegie Mellon University
Pittsburgh, PA 15213-3890

Some multiple stage optimization models are discussed at the end of the first chapter.

The second chapter presents a new model for portfolio selection with order of expected returns. In the new model, the expected returns of securities are put as variables rather than as the arithmetic means of securities. Due to the non-concavity and special structure of the model, the new model is solved by a particularly designed genetic algorithm. Compared with the performance of the traditional model of Markowitz when arithmetic means of securities are directly put as the expected returns of securities, the new model tends to have greater returns and variance.

Chapter 3 proposes a compromise solution to mutual funds portfolio selection with transaction costs. The transaction costs are assumed to be a V-shaped function of the difference between an existing portfolio and a new one. Under some reasonable assumptions on the variance-covariance matrix of returns, a compromise solution to the portfolio selection problem is found. The compromise solution has four good properties which are respectively minimal regret, Pareto efficiency, scale independence and no dominance of one criterion over another. Also the compromise solution is extended to the case including a riskless asset.

The fourth chapter studies an optimal portfolio selection problem for assets. A general model is formulated by taking into consideration transaction costs and taxes. Short sales and borrowing are not allowed in the model. Some properties of the efficient portfolio and efficient frontier are established. An interactive method which requires only paired preferences of the investor is also presented.

Chapter five considers all kinds of portfolio selection problems with different interest rates for borrowing and lending in which the riskfree borrowing rate is higher than the riskfree lending rate. Different efficient portfolio frontiers are derived and graphed in this chapter.

The sixth chapter illustrates some results of the stochastic programming methods for solving the multi-period portfolio selection. After a brief review of models of multi-period investment problems, some empirical results of both discrete and continuous models are given.

The seventh chapter proposes a mean-variance-skewness model for portfolio selection with transaction costs. By a transformation, the non-convex and non-smooth programming problem is converted to a linear programming problem. This implies that the model can be used efficiently to solve large-scale investment problems for mutual funds or any financial institutions in a world wide financial market.

The eighth chapter illustrates the capital asset pricing theory. The standard form of CAPM is explained and all kinds of non-standard derivatives are explained. The multi-period inter-temporal capital asset pricing model and the arbitrage pricing theory are also illustrated.

The ninth chapter tests the two-parameter version of the asset pricing model for the stock markets of China which include both type A stocks and type B stocks listed in the Shanghai Stock Exchange and the Shenzhen Stock Exchange. Empirical results of the CAPM with and without considering taxes on stock dividends are presented. Results show that China Stock Markets do not support the linear relations of CAPM.

Several chapters of this book are based on the joint work of the two authors during 1996~1999, at which time the second author pursued his MSC in the Institute of Systems Science, Chinese Academy of Sciences. The second author greatly appreciates the first author's supervision and help.

We have to thank a number of people for their help and support in our preparing this book. First, we thank Prof. Xiaotie Deng of City University of Hong Kong for his joint work on revision of three chapters of this book. We also thank Prof. Sunming Zhang of Tsinghua University in Beijing and Prof. Zhongfei Li of Zhongshan University and Prof. Shancun Liu at Beijing Aerospace and Aeronautics University for their contribution to this book. Three chapters are based on the results in which the first author achieved jointly with them. We thank a number of scientists for helpful discussions and valuable comments, among them are Prof. T. Tanino of Osaka University, Prof. Duan Li of Chinese University of Hong Kong, Prof. Y.Yamamoto of University of Tsukuba, Prof. Y Nakamaya of Konan University in Kobe, Prof. Jianming Shi of Tokyo University of Science and Prof. Zaifu Yang of National Yakohama University. We thank Mr. Chen Ji for his patience in typing the manuscript of this book. Chuck Peek at Austin has been great helpful in revising the book.

Finally, we would like to thank National Natural Science Foundation of China (NSFC), Chinese Academy of Sciences, Hong Kong Research Granting Committee, University of Texas at Austin in USA and University of Tsukuba in Japan for their financial support.

Shouyang Wang
Institute of Systems Science
Academy of Mathematics and Systems Sciences
Chinese Academy of Sciences
Beijing, 100080, China

Yusen Xia
Red McCombs School of Business
University of Texas at Austin
Austin, TX 78712, USA

December, 2001

Contents

1. Criteria, Models and Strategies in Portfolio Selection

1.1 Introduction

The uncertainty about future events makes the behavior of economic agents unpredictable and, at times, brings about turbulence to financial markets. Assumptions about their behavior in resource allocation under an uncertain and ever-changing environment make up the foundation for theories of economics and finance to develop. These theories have thus gone a long way to apply mathematical analytical tools to model both the behavior of economic agents and future events of financial markets. Resource allocation methods derived from mathematical models of modern finance in turn play an influential role in the practice of financial institutions, and in a way, become a non-negligible player in the game of financial markets.

It has been suggested that the origin of modern mathematical models in finance can be traced back to Louis Bachelier's dissertation on the theory of speculation [Cootner (1989), Howison *et al.* (1995), Mandelbrotm (1989)]. However, without doubt, the earth-breaking work of Markowitz in portfolio selection [Markowitz (1952)] has been most influential for the development of modern mathematical finance and its applications in practice. The Markowitz theory of portfolio management deals with individual agents in the financial markets. It combines probability theory and optimization theory to model the behavior of economic agents. The agents are assumed to strike a balance between maximizing the return and minimizing the risk of their investment decision. Return is quantified as the mean, and risk as the variance, of their portfolio of securities [Markowitz (1952,1959)]. These mathematical representations of return and risk have allowed optimization tools to be applied to the studies of portfolio management. The two objectives of investors of profit maximization and risk minimization are thus quantified as to maximize the expected value and to minimize the variance of their portfolio. The exact solution will depend on the level of the risk (in comparison with the rate of the return) that the investors would bear. Even though many later models may have different point of view on what mathematical definitions should be given for risk and return of economic agents, the trade-off between return and risk (maybe in different forms) has been the major issue which those theories try to solve.

In this chapter, we will emphasize on how mathematical theories about return and risk are applied to develop criteria, models and strategies for portfolio management for economic agents. Therefore, our main concern will be individual agents in a financial market. Hence we will have to leave out many interesting developments in mathematical finance. This includes, for example, the Capital Asset Pricing Model (CAPM) [Sharpe (1964), Lintner (1965), Mossin (1966) and Fama (1997)], the Arbitrage Pricing Theory (APT) [Ross (1976,1977)] and the option pricing theory [Hull (1993)].

In Section 1.2, we discuss the Markowitz model and its modifications as well as the related models. In Section 1.3, we discuss models which are based on different criteria for risk and return but share the same feature as the Markowitz model that there is an underlying probability distribution for changes in the stock market. Some are more closely related to the Markowitz model than others. In Section 1.4, we consider models in which a decision does not rely on probability distributions on stock movement though such information may still be utilized. When no probability distribution is known, one may seek a solution which is the best in the worst case. When it is known, we may find a solution which is the best in the average. In Section 1.5, we discuss the multi-stage optimization models. We conclude this chapter with a few remarks and discussions in Section 1.6.

1.2 Mean and Variance as Return and Risk

In the Markowitz model, a probability distribution of security prices is assumed to be known, and the return of any portfolio is quantified as its expected value and its risk is quantified as its variance. It appears that Markowitz simply presented his model with mean and variance as return and risk. However, this is not done without justification. Markowitz discussed the principles that justify such a choice [Markowitz (1952)]. In general, the introduction of the variance is presented as a quadratic approximation to a general utility function. Interesting examples of utility functions were analyzed to justify this approach [Markowitz (1952)].

1.2.1 Markowitz formulation of portfolio optimization

In a standard formulation of Markowitz model, an investor will choose x_i, the proportion invested in security i, $1 \le i \le N$ for N securities. The constraints are

$\sum_{i=1}^{n} x_i = 1$, and $\forall i: x_i \geq 0$. The return R_i for the $i-th$ security, $1 \leq i \leq N$, is a random variable, with expected return $\mu_i = E(R_i)$. Let $R = (R_1, R_2, \cdots, R_N)^T$, $x = (x_1, x_2, \cdots, x_N)^T$ and $\mu = (\mu_1, \mu_2, \cdots, \mu_N)^T$. The return for the portfolio is thus $R^T x = \sum_{i=1}^{N} R_i x_i$ with expected return $E(R^T x) = \mu^T x = \sum_{i=1}^{N} x_i \mu_i$. Let Σ be the variance-covariance matrix of random vector R. Therefore, the variance of the portfolio is $V = x^T \Sigma x$. Investors are assumed to like return and dislike risk. Thus, given a fixed risk level, the investors will choose a portfolio which maximizes return. On the other hand, for a given level of return, the investors will choose a portfolio of the minimum risk. This is called the mean-variance efficiency. A portfolio satisfying this condition is called an efficient portfolio. More specifically, a portfolio is inefficient if there is a portfolio, either with larger mean and no larger variance, or with smaller variance with no less mean. Otherwise, the portfolio is efficient. The set of all the efficient portfolios is called the efficient frontier. Taking into consideration that each investor may weigh risk and return in different ways, no attempt is made to spell out the exact preference of every individual in the Markowitz model. However, their choices must be in the efficient frontier

Under this formulation, the portfolio management problem becomes a mathematical programming problem with two objectives: maximizing the mean and minimizing the variance. It is possible to have different ways to solve this problem, depending on the preference of investors. As suggested above, two simple cases are: maximizing return given a fixed level of risk, and minimizing risk given a fixed level of return.

Let γ be the maximum risk level the investor would bear. The mathematical programming formulation for the former is as follows:

$$\text{maximize } E(R^T x)$$

$$\text{subject to } (x^T \Sigma x)^{\frac{1}{2}} \leq \gamma$$

$$\sum_{i=1}^{n} x_i = 1$$

$$x_i \geq 0, \ i = 1, \cdots, N$$

Σ can be shown to be positive definite under the assumptions that all the assets do not have the same return and that they are linearly independent [Green (1986), Mcentire (1984)]. In general, Σ is positive semi-definite because it is the variance-covariance matrix of random vector R.

Notice that

$$\forall x:\ x^T \sum x = \sum_{i,j} x_i (E[X_i X_j] - E[X_i]E[X_j]])x_j$$

$$= E[(\sum(x_i X_i - E[x_i X_i])^2] \geq 0$$

The inequality $(x^T \sum x)^{\frac{1}{2}} \leq \gamma$ (or equivalently $x^T \sum x \leq \gamma^2$) defines a convex set of variables x. All other constraints are linear in x. This is thus a problem of optimizing a linear function over a convex set. Moreover, given a point x^0 in the space, it is easy to check whether it is a feasible solution to the above constraints. If not, a hyperplane separating x^0 from the above convex set can easily be found in polynomial time. This is straightforward for the linear constraints and can be set to be the hyperplane tangent to the ellipsoid $x^T \sum x \leq \gamma^2$ for this quadratic constraint. It can be solved in polynomial time by the ellipsoid method [Martin *et al.* (1987)].

In the case of minimizing risk for a given level of return, the mathematical programming formulation is

$$\text{minimize } x^T \sum x$$

$$\text{subject to } E(R^T x) \geq \alpha$$

$$\sum_{i=1}^{n} x_i = 1$$

$$x_i \geq 0,\ i = 1, \cdots, N$$

where α represents the minimum return the investor would accept. Since $x^T \sum x$ is a quadratic convex function, the problem can be again solved in polynomial time.

1.2.2 Supporting evidence for Markowitz model

Markowitz noticed that not all utility functions can be accurately approximated by a quadratic curve as in the mean-variance analysis. On the other hand, he also pointed out that it is surprisingly flexible in approximating smooth concave curves [Kroll *et al.* (1984), Markowitz (1959,1987)]. After a careful discussion of various studies on approximation of utility functions using the mean-variance analysis, Markowitz concluded that it provides almost maximum expected utility except for those with pathological risk aversion [Markowitz (1987)].

One interesting and related work is that of Kallberg and Ziemba who, after examining several different utility functions, showed that similar portfolios can be obtained using those functions if they have the same Arrow's absolute risk aversion coefficient, see [Kallberg and Ziemba (1983)].

In an empirical study of Markowitz model, Farrar (1965) attempted to compare its prediction with actual portfolios held by mutual funds and showed that it is a quite good predictor of the actual behavior.

Farrar's main conclusions are as follows:

The characteristics of the mutual funds' portfolio are quite similar to those of corresponding efficient portfolios.

Mutual funds claimed to be risky have portfolios corresponding closely to risky efficient portfolios.

Balanced funds correspond more closely with low risk efficient portfolios.

Perold (1984) observed that the mean-variance model gained widespread acceptance as a practical tool for portfolio selection. Perold found that mean-variance efficient portfolios had been routinely computed as part of the portfolio allocation process by many investment advisory firms and pension plan sponsors. In particular, the methodology has been applied to asset allocation, multiple money management decisions, index matching and active portfolio management.

1.2.3 Index models

Although we have seen that the basic portfolio optimization problem in the mean-variance model can be solved in polynomial time, historically, there has been a major effort spent to reduce the computation requirement. The single index model of Sharpe (1963) is an early breakthrough in this direction. Notice that, for N securities, there are $N*(N+1)/2$ covariance coefficients in the variance-covariance matrix Σ. At the time when high performance computers were not available, it was computationally burdensome to estimate Σ even for moderately many securities. Sharpe made the observation that the prices of most stocks increase when the market goes up and decrease when it goes down. A market factor was introduced to describe this type of movements of securities. Differences between the returns of individual securities were assumed to be the result of additional independent random disturbance specific to each security.

More formally, the return of a security is broken into two parts. One depends on the market, and the other is a random variable independent of those of other securities:

$$R_i = \alpha_i + \beta_i R_m + e_i$$

R_m is a random variable representing the rate of return due to a market index. The constant β_i measures the market's effect on return R_i of security i (*i.e.*, its sensitivity to market movements). The constant α_i is a component (independent of the market) for return R_i of security i. The expected value of random variable e_i is zero $(E(e_i)=0)$ with variance σ_{e_i}. It is assumed that R_m and e_i are independent. That is, $\operatorname{cov}(R_m, e_i)=0$. Another important assumption is that $\{e_i : i=1,\cdots,N\}$ are jointly independent. This means that the sole reason that stocks vary together is due to market fluctuations R_m. The expected return and variance of the portfolio are given by:

$$E(R_p)=E(R^T x)=E((\alpha_i+\beta_i R_m)^T x)=\sum_{i=1}^{N} x_i(\alpha_i+\beta_i E(R_m))$$

and

$$\sigma_p^2=\sum_{i=1}^{N} x_i^2\beta_i^2\sigma_m^2+\sum_{i=1}^{N} x_i^2\sigma_{e_i}^2+\sum_{i=1}^{N}\sum_{j=1,i\neq j}^{N} x_i x_j\beta_i\beta_j\sigma_m^2$$

One may observe that the single index model reduces the estimation of variance-covariance coefficients from $O(N^2)$ to $O(N)$.

In addition to the market movement, other factors such as industry effects are observed to have influence on security prices [King (1966)]. Believing that this may improve upon the single index model, several multi-index models have been suggested. Those models have used industrial classifications [Cohen and Pogue (1967)], tendency of firms to act alike [Elton and Gruber (1970)], and statistical procedures to determine pseudo-industries [Farrell (1974)]. It has been observed [Elton and Gruber (1971)] that additional indices led to a decrease in performance. In comparison with the multi-index models, the single index model leads to lower expected risk and at the same time was much simpler to use [Cohen and Pogue (1967), Elton and Gruber (1973), Elton *et al.* (1978)].

1.2.4 Mean-semivariance model - (E-S)

Based on the observation that investors may only be concerned with the risk of return being lower than mean (downside risk), the method of mean-semivariance (E-S) was proposed to model this fact (see, e.g., [Markowitz (1959), Mao (1970a, 1970b) and Swalm (1966)]). The semivariance is defined as the expected value of squared "positive (or negative)" deviations from the mean (or more generally a value chosen by the investor as a critical value). More formally, if we let R be the random variable for the price of a security, the semi-variance S is defined as $E[(\min\{R-E(R),0\})^2]$.

Although the rational behind introducing the model of mean-semivariance is intuitively closer to reality than the mean-variance model, it is not widely used. First of all, people are usually more familiar with techniques for analysis of variance developed in statistics than semi-variance. Secondly, when the probability distribution is symmetrical, semi-variance is proportional to variance. Finally, both the mean variance model and the mean semi-variance model assume that the investor has an increasing absolute risk averse utility function which limits its generality [Bawa (1975)].

1.2.5 Mean absolute deviation model - (MAD)

In order to solve large-scale portfolio optimization problems, Konno and Yamazaki (1991) considered the mean-absolute deviation as the risk of portfolio investment. The risk of absolute deviation is defined as

$$w(x) = E[\left|\sum_{j=1}^{N} R_j x_j - E[\sum_{j=1}^{N} R_j x_j]\right|].$$

It is obviously an L_1 metric version of the mean-variance approach.

Under the context of minimizing risk given a specified return value, or maximizing the return given the mean absolute deviation, Konno and Yamazaki derived a linear program for portfolio selection as follows:

$$\text{minimize } \frac{1}{T}\sum_{t=1}^{T} y_t$$

$$\text{subject to } y_t + \sum_{j=1}^{n} a_{jt} x_j \geq 0,\ t = 1, \cdots, T$$

$$\sum_{j=1}^{n} r_j x_j \geq \rho M_0$$

$$\sum_{j=1}^{n} x_j = M_0$$

$$0 \leq x_j \leq \mu_j,\ j = 1, \cdots, n$$

Using the historical data of Tokyo Stock Exchange, Konno and Yamazaki compared the performance of the mean variance model and the mean absolute deviation model and found that the performance of those two models were very similar.

Feinstein and Thapa (1993) presented a reformulation of the mean absolute deviation model that is equivalent to the model of Konno and Yamazaki and at the same time reduces the bound on the number of non-zero assets in the optimal portfolio. While Konno and Yamazaki showed that the mean absolute deviation model did not require the covariance matrix, Simaan (1997) found that this would result in greater estimation risk.

1.2.6 Mean target model (MTM)

Fishburn (1977) examined a class of mean-risk dominance models in which risk equals the expected value of a function that is zero at or above a target return t and is non-decreasing in deviations below t. Considering that investment managers frequently associate risk with the failure to attain a target return, the $\alpha - t$ model is formulated as follows:

$$\text{maximize } F_{\alpha}(t) = \int_{-\infty}^{t} (t-x)^{\alpha} dF(x), \ \alpha > 0$$

where t is a specified target return, x is the return of investment, α is a parametric which can approximate a wide variety of attitudes towards the risk of falling below the target return. F is the probability distribution function of portfolio return.

The $\alpha - t$ model avoids distributions having below-target returns. However, a major drawback of the computational complexity restricts this model's ample uses in practice.

1.2.7 Mean-variance-skewness model (MVS)

The third moment of a return distribution is called skewness which measures the asymmetry of the probability distribution. A natural extension of the mean-variance model is to add the skewness in consideration for portfolio management. There will be triple goals: maximizing the mean and the skewness, minimizing the variance. People interested in the use of skewness prefer a portfolio with a larger probability for large payoffs when mean and variance remain the same. The mathematical programming formulation is as follows:

$$\begin{aligned} \text{maximize } & E[(R(x)-r(x))^3] \\ \text{subject to } & E[(R(x)-r(x))^{2]} = s^2 \\ & r(x) = r \\ & x \in X \end{aligned}$$

where $X = \{x \in R^n : Ax = b, x \geq 0\}$, r is the expected return and s^2 is the variance of candidate portfolios.

The importance of higher order moments in portfolio selection has been suggested by Samuelson (1958). But partially because of the difficulty to estimate the third order moment for a large number of securities such as over a few hundreds and to solve the non-convex programs by using traditional computational methods, quantitative treatments of the third order moment have been neglected for a long time.

Since high performance computers are becoming cheaper, it is conceivable that the use of skewness in portfolio analysis would become reality in the near future. The main question here is whether introducing the skewness would significantly improve quality of chosen portfolios. Starting in the 1990s, several quantitative results have appeared to study the optimal portfolio when the skewness is taken into consideration. For example, Konno and Suzuki (1995) applied piecewise linear approximation to obtain solutions in this model. Very recently, Liu, Wang and Qiu (2000) presented a mean-variance-skewness model for portfolio selection with transaction costs and proposed an algorithm which can solve the model efficiently.

Chunhachinda *et al.* (1997), based on their result that the return of the major stock markets all over the world are not normally distributed, showed that taking skewness into consideration would result in a major change in the optimal portfolio. But whether this will benefit the investors is still unknown.

1.2.8 Multi-objective optimization methods

One may view the Markowitz model as a bi-objective optimization program: maximizing return and minimizing risk. Actually, a few multi-objective optimization techniques can be applied to the study of portfolio selection.

Levy and Markowitz (1979) considered functions of two objectives which are respectively mean and variance and used Taylor extension deriving the result that any utility function can be approximated by functions of mean and variance.

The methodology of goal programming [Kumar *et al.* (1978), Lee and Lerro (1973), Tamiz (1996)] has been designed to resolve situations where a single objective optimization assumption is replaced by the more realistic case that several conflicting goals may compete in the allocation decision. Even though the Markowitz mean-variance model is inherently bi-objective, solutions have been sought with techniques of scale optimization. The technique of goal programming is best suited for problems where investors seek two goals satisfying rather than optimizing. Here both goals and constraints are incorporated in the allocation decision. An objective function is constructed such that its solution is as close as possible to both goals.

In suggesting multi-objective programming techniques for portfolio selection, Tamiz (1996) proposed a two-stage model in which sensitivity of shares to specific macro-economic factors is determined in the first stage and a portfolio based on the investor's preferences is selected in the second stage. Ballestero and Romero (1996) considered compromise programming solutions in the portfolio selection. Their solution applies graphic techniques to determine the compromise set. They found that the compromise solution was surrogate to average investors' utility maximization. Li, Wang and Deng (2000) considered a bi-objective model for portfolio selection with transaction costs and they also derived a linear programming problem by some transformations.

Xia, Liu, Wang and Lai (2000) considered a bi-objective portfolio selection model with the order of expected returns of securities and compared their models with the traditional Markowitz mean-variance model. Liu, Wang and Qiu (2000c) proposed a three objective programming model for portfolio selection with transaction costs. Shing and Nagasawa (1999) proposed an interactive multiple criteria decision making model for portfolio selection. The model considered both the return and the risk of the portfolio.

1.3 Different Criteria for Risk and Return

In a more general setting, in comparison with the mean-variance analysis, the investor's problem is to maximize the expected utility E[$U(R)$] of of an investor, which is a more accurate portrayal of the investor's preference in seeking return and bearing risk. When the return is a random variable, this leads to the following stochastic optimization problem:

$$\text{maximize } E_r(U(R^T x))$$

$$\text{subject to } \sum_{i=1}^{n} x_i = 1$$

Several commonly accepted assumptions are as follows:

1. The investor maximizes the expected utility u ;
2. the investor prefers more to less;
3. the investor is risk averse, *i.e.*, for any random return R_i, she/he prefers sure payment $E(R_i)$ to the random payoff R_i ;
4. the utility function is continuous, monotonely increasing, concave: $U \in C_1$; $U' > 0$; $U'' < 0$; $|U'|, |U''| < \infty$.
5. an optimal solution x^* exists, R and the variance-covariance matrix Σ has finite entries.

Obviously, the mean-variance approach to portfolio selection is accurate only for normal distributions of returns or quadratic utility functions. Although it can be a good approximation of the reality for the situations where these conditions fail to hold, different models have been suggested with different criteria for risk and return. In this section, we discuss several models based on a probability assumption on security price changes.

1.3.1 Baumol's confidence limit criterion--(E-L)

Baumol (1963) suggested that an investor would usually not be interested in the entire efficient frontier of portfolios yielded by the Markowitz mean-variance model but is often concerned with a smaller subset of efficient portfolios. According to Baumol, investors are not only concerned with obtaining an expected future return while minimizing their risks. Instead, they may concern with the minimum acceptable return. More formally, given two investments, F and G, F will be preferred to G if

$$E_F \geq E_G \text{ and } E_F - K\delta_F \geq E_G - K\delta_G$$

where $E_F(E_G)$ is the expected return for portfolio F (G) and δ_F (δ_G) the standard deviation. The constant K represents the investor's coefficient of risk aversion and the expression $E - K\delta$ represents the "lower confidence limit for the investor's return". The critical nature of the risk-aversion coefficient is defined by the Tchebycheff inequality, which states that

$$\Pr[|F - E_F| > K\delta_F] \leq \frac{1}{K^2}$$

That is, the probability that the actual return from a portfolio F will be K standard deviations below the expected return E_F of F is bounded by $\frac{1}{K^2}$.

Thus, Baumol held that investors might construct a (smaller) set of efficient portfolios based on expected gain, E, and a lower confidence limit, L. It can be easily shown that Baumol's E-L criterion yields an efficient set which is a subset of the Markowitz mean-variance efficient portfolios. In fact as $K \to \infty$, the E-L frontier approaches the set of mean-variance efficient portfolios [Russell and Smith (1966)].

1.3.2 Minimizing probability of specified risk

While Markowitz's mean-variance model measures risk with the variance of a portfolio, Roy (1952) proposed another criterion to minimize the probability that a portfolio reaches below a disaster level. Along the similar line of the approach is to maximize the probability of achieving certain investment goals [Williams (1997)]. Denoting by R_p the return on the portfolio p and by D the minimum acceptable level of return for the investment, Roy's criterion is

$$\text{minimize } \Pr(R_p < D)$$

Here the risk is described facing the situation where the outcome of one's investment drops below the accepted level D. The concern of the investor is to minimize the probability of this event. With this different view of risk (suggested by Roy (1952)) contemporaneously with the view of Markowitz of variance as risk), the optimization problem can have several different formulations [Elton and Gruber (1995)]. One example is the Kataoka criterion: maximize the lower limit subject to the constraint that the probability that, a return is less than, or equal to, the lower limit D, is not bigger than some predetermined value. Let α be the predetermined probability the investor may tolerate for the investment to drop below the limit D. This criterion can be stated as follows:

$$\text{maximize } D$$

$$\text{subject to } \Pr(R_p < D) \le \alpha$$

The Telser criterion, on the other hand, maximizes expected return, subject to the constraint that the probability that, a return less than, or equal to, some predetermined limit D, is not bigger than some predetermined number α. In mathematical form, this is stated as follows:

$$\text{maximize } E(R_p)$$

$$\text{subject to } \Pr(R_p \le D) \le \alpha$$

Along a similar line approach is to maximizing the probability of achieving certain investment goals [Williams (1997)].

1.3.3 Three stochastic dominance criteria

In deriving the mean-variance model, Markowitz started from some basic axioms that formulate investors' preference to derive a utility function (and its quadratic approximation). The Stochastic Dominance (S-D) Criteria [Hadar and Russlell (1969), Quirk and Saposnik (1962), Whitemore (1970)], however, focus on fundamental principles that determine behavior of investors. Three related criteria were proposed. One refines another. The first order stochastic dominance states that investors prefer more to less, as the second order, one adds to the first the assumption that they are risk averse, and the third adds one more assumption that they have decreasing absolute risk aversion.

More formally, let $f(R)$ and $g(R)$ represent the probability density functions of returns from portfolios F and G, respectively, with domain $[a,b]$.

The First Order Stochastic Dominance (FSD) Criterion [Hadar and Russlell (1969), Quirk and Saposnik (1962)] states that portfolio F dominates portfolio G if

$$\int_a^R [f(y) - g(y)]dy \le 0 ,$$

for each $R \in [a,b]$, with a strict inequality for at least one $R \in [a,b]$. The only specification about the investors' utility function necessary for the FSD criterion is that it is a monotonically increasing function, *i.e.*, $U(R) > 0, \forall R \in [a,b]$.

The Second Order Stochastic Dominance (SSD) Criterion [Hadar and Russlell (1969), Hanoch and Levy (1969), Hanoch and Levy (1969,1970)] states that portfolio F dominates portfolio G if

$$\int_a^R \int_a^y [f(w) - g(w)]dwdy \le 0$$

for all values of $R \in [a,b]$, with a strict inequality for at least one $R \in [a,b]$. The specification for the decision-maker's utility function is more restrictive than that for FSD. In the SSD case, the utility function must have $U'(R) > 0$ and $U''(R) < 0$, $\forall R \in [a,b]$.

The Third Order Stochastic Dominance (TSD) Criterion [Levy and Sarnat (1970), Philippatos and Gressis (1975), Porter (1973)] states that portfolio F dominates portfolio G if

$$\int_a^R \int_a^y \int_a^w [f(z) - g(z)]dwdydz \le 0$$

for each $R \in [a,b]$, with a strict inequality for at least one $R \in [a,b]$, and

$$\int_a^R \int_a^y [f(w) - g(w)]dwdy \le 0$$

where R varies continuously on the closed interval $[a,b]$. The TSD is appropriate for situations where we assume a decreasing premium associated with an investor's wealth. In such cases the investor's utility function must have $U'(R) > 0$, $U''(R) < 0$ and $U'''(R) > 0$, $\forall R \in [a,b]$.

1.3.4 Entropy as a measure of risk

As an alternative measure of uncertainty for portfolios, Philippatos and Wilson (1972, 1974) proposed that the use of entropy (or expected information) would be more dynamic and general as well as free from reliance on specific distributions. Here uncertainty is used as a synonym for risk in the sense that investors dislike uncertainty. Therefore, the dual goals for portfolio selection in this model are to maximize the return and minimize the entropy.

The entropy for a portfolio with density function $f(x)$ is $\int f(x)\log f(x)dx$. For some probability distributions, e.g., the uniform distribution and the normal distribution, it can be shown that entropy and variance are monotone functions of each other. To calculate the entropy of a portfolio, the joint entropy of the constituent stocks needs to be evaluated. This can be a forbidden task when the number of stocks is moderate. Philippatos and Wilson (1972) chose to combine this approach with that of single index model of Sharpe. In their model, for a portfolio $P = \sum_{i=1}^{N} x_i R_i$, its entropy (conditioning on market portfolio R_I) is thus

$$H(P|R_I) = \sum_{i=1}^{N} x_i H(R_i|R_I)$$

This allows them to test the scheme of mean-entropy diversification in portfolio selection. Using monthly relative returns of a sample of 50 randomly selected securities for the period from 1957 to 1971, they showed that the efficient portfolios of the mean-entropy model were consistent with those of the mean-variance model and the single index model. Liu, Wang and Qiu (2000b) presented a model for mutual funds to choose portfolios based on entropy.

1.3.5 Maximizing the geometric mean return

The geometric mean of a random variable R, suggested as an alternative criterion for portfolio selection [Latane (1959)], is defined as follows:

$$\prod_{i=1}^{N}(1+r_i)^{p_i} - 1$$

where the probability of the random variable R assuming the value r_i is p_i. One of the convincing evidence for the Markowitz model is its prediction power for a diversified portfolio. This is also shown for the method of geometric means. The portfolio with maximum geometric mean is often diversified. If returns are log-normally distributed, the portfolio that maximizes the geometric mean return is

mean-variance efficient. In general, Latane (1959, 1967, 1969) showed that the portfolio with the highest geometric mean return would have the highest expected value of terminal wealth. It can also be shown that it has the highest probability of exceeding any given wealth level over any given time period [Brieman (1960)]. Notice that this criterion has no mention of the utility function of investors. Believers of utility theory quickly found out that maximizing the expected value of terminal wealth was not identical to maximizing the utility of terminal wealth. They considered this a major flaw of the geometric mean method.

1.4 Best in the Worst Case or in the Average

Even though it is common to formulate uncertainty as probability distribution, we have to point out that this is not the only principle for dealing with uncertainty.

In the following we introduce several approaches different from a probability distribution that are suggested to deal with unknown future events. In case when probability is indeed unknown, we may choose to find a strategy which is the best for the worst case. If we are lucky to have a probability distribution, we may find the best strategy in the average.

1.4.1 Competitive analysis

The competitive analysis method takes into consideration that the optimal solution may not be achievable in the presence of uncertainty. It aims at a solution consistent with available information which is an approximation of the optimal solution for each possible outcome. The worst case ratio of this solution to that of optimum is called the competitive ratio.

To illustrate the competitive analysis method, consider the special case when we know no information about future prices of the stocks, similar to the assumption for Cover's universal portfolio [Cover (1991)]. We would analyze the strategy of distributing the fund equally to all the stocks such that $b_1 = b_2 = \cdots = b_m = \frac{1}{m}$. Notice that this is the same as the portfolio in the first stage of the universal portfolio. Let x_i be the relative price of stock i at the end of the period, *i.e.*, its price at the end of the period divided by its price at the beginning of the period. The outcome of the above simple strategy is $\sum_{i=1}^{m} \frac{x_i}{m}$.

On the other hand, if future information is available, a wise investor would have put all the money in the stock of the best performance to achieve a rate of increase of x_k =max $\{x_i : 1 \le i \le m\}$. Thus, the competitive ratio is

$$\frac{\sum_{i=1}^{m} \frac{x_i}{m}}{x_k} \ge \frac{1}{m}.$$

That is, this simple strategy achieves a competitive ratio of $\frac{1}{m}$. Even though this does not look very good, one may show that, no strategy can achieve a competitive ratio better than $\frac{1}{m}$ if no information about the securities is available [Deng (1998)].

For the competitive ratio, all the possible outcomes consistent with the available information are considered in the worst case ratio. Dependent on what kind of information is available to the investors, it may result in different computational procedures (see [Bell and Cover (1980), Deng (1996, 1998)]). A similar problem [Chou (1995)] is considered under the assumption that a statistical adversary determines the future prices of a certain security (foreign currency). Wang (2000) made some remarks on the framework of competitive analysis and expected some other types of efficient approaches to solve online portfolio optimization problems.

1.4.2 Scenario analysis

Scenario analysis was proposed by Rockafellar and Wets (1991) to deal with uncertain future where only partial or no information is known. Here investors specify a small number (obviously, to reduce computational cost) of possible scenarios, each corresponds to a possible future outcome with an assigned probability value. The goal is to choose decision variables to optimize the expected value of objective function values of this collection of optimization problems.

To illustrate the main ideas of scenario analysis, we consider a simplified version of linear programming as presented in [Dembo (1989,1990)] as follows:

$$\begin{aligned} \text{maximize } & c_s^T x \\ \text{subject to } & Ax = b \\ & x \geq 0 \end{aligned}$$

where c_s is an uncertain vector dependent on the scenario s. Denote the optimal solution value by v_s for the above linear program.

The problem we actually face is that we do not know the value of c_s when decision x is made. In scenario analysis, we are interested in a solution x such that $c_s^T x$ is as close to v_s as possible for each scenario s, in a way very similar to coordination theory for multi-agent decision making in economics [Bicchieri (1993)]. The closeness is defined by the metric chosen by the investor. It results in the following mathematical program if the Euclidean distance is chosen:

$$\text{maximize } \sum_s \left\| c_s^T x - v_s \right\|^2$$

$$\text{subject to } Ax = b$$
$$x \geq 0$$

If the probability p_s of each scenario is known, we can use the weight objective function $\sum_s p_s \left\| c_s^T x - v_s \right\|^2$.

It is noticed [Dembo (1989)] that immunization portfolios are extensively used by some investors, especially mutual fund managers. Even though we present this method as if it is for a single period problem, it has been originally proposed as a multi-stage stochastic programming technique (see [Dembo (1990), Robinson (1991), Sengupta (1986)]).

1.4.3 Tracking models

While both competitive analysis and scenario analysis compare the performance of a strategy with that of the optimal solution (unachievable because of unknown information), the tracking models[Clarke *et al.* (1994), Dembo and King (1992)] evaluate their performance with some benchmark portfolio, e.g., that of some diversified index of assets.

As a simple model suggested by Dembo and King (1992), the following optimization problem tracks the target by minimizing the expected deviation of the random portfolio return from the random target return:

$$\text{maximize } \sum p_s [(r^s)^T x - v_s^*]_-^2$$
$$\text{subject to } e^T x \leq 1$$
$$0 \leq x \leq 1$$

where v_s^* is the scenario-optimal value. We should notice that one may use other metric to evaluate the difference from the optimal value v_s^*.

One may notice that the net effect of this tracking problem is to penalize the sum of the squared deviations of $(r^s)^T x - v_s^*$ over those scenarios where this difference falls below zero. That is, in a way very similar to the idea behind mean-semivariance approach, it models the downside risk of that does not meet the target v_s^*.

1.4.4 Two person game models

A portfolio optimization problem can be viewed as a game between an investor and nature, which makes its choice versus the best decision the investor may make. Dependent on our view of nature, the performance metric ranges from the worst case ratio in comparison with the optimum for competitive analysis, to an average in scenario analysis, and then to an average comparison with an index portfolio for tracking models. Whether we make any probability assumption or not in the above models, the essence of the analyses is to design a solution which is acceptable in comparison with certain solutions perceived as good ones by investors. This has already put our discussion in the vicinity of the reign of game theory methods such as coordination theory.

Sengupta (1989a) viewed the portfolio decision problem as a two person game where the first player was an investor and the second player as the overall market which provided signals for the first player to choose his/her strategies. One may consider a single stage play or multiple plays in the stock market. Sengupta formulated the two-person zero-sum game for portfolio selection as follows:

$$\max_{x \in R_x} \min_{y \in R_y} x^T By$$

where $B = (r_{ij})$ is the matrix for returns, $R_x = \{x | \sum_{i=1}^{m} x_i = 1, x_i \geq 0; i = 1,2,\cdots,m\}$ represents the set of decisions for the investor, and $R_y = \{y | \sum_{i=1}^{n} y_i = 1, y_i \geq 0; i = 1,2,\cdots n\}$ is the set of choices for the market as an adversary. This results in a formulation very similar to that of competitive analysis. In the case of zero-sum game, Sengupta characterized mixed strategies in different ways for both static and dynamic games. It is also shown that the solution for the mean-variance model could be generalized [Sengupta (1989a)]. One may find several other interesting portfolio selection methods using game theory approach in [Bell and Cover (1988), Sengupta (1989b), Wang (2000), and Deng, Li and Wang (2000)].

1.5 Multiple Stage Models

The models in the previous sections are mainly concerned with one period portfolio optimization. In reality, however, the decision problem of investors is one of multiple stages. They change their portfolio holdings from time to time aiming to maximize their utility functions. Multi-stage and dynamic models have been suggested as more realistic for portfolio optimization of investors (for example, see, [Elton and Gruber (1974), Fama (1970), Hakansson (1971), Maclean and Ziemba (1991), Roll (1973), Samuleson (1983), Sengupta (1983)]). However, for quite sometime those approaches have mainly been theoretical due to the computational difficulty. As the development of high performance computers, more and more research efforts have been made in this direction.

1.5.1 Capital growth theory

Noticing that the maximum of expected compounded return ($Ex[T^{1/n}]$) is not the same as the maximum of compounded expected return ($Ex[T]^{1/n}$) in an n-period investment problem with T representing the terminal wealth, Hakansson (1971, 1974) suggested that it might not be adequate trying to solve a multiple stage investment problem by a sequence of one stage Markowitz mean-variance solutions. Even though the capital growth model is not based on any utility function, it is consistent with logarithmic utility functions and also fits the data well [Roll (1973)]. The model is also considered consistent with the capital asset pricing model [Merton (1973a), Michaud (1981), Roll (1973)]. MacLean *et al.* (1992) applied the capital growth idea to the analysis of dynamic investment and Wilfox (1998) studied its power in the investment of an emerging market that is characterized with high returns as well as high risk. Li and Ng (2000) formulated a dynamic programming model to derive an analytic solution to a multi-stage portfolio selection problem. Li, Li and Wang (2000a) got a similar derivation for multi-period efficient portfolio frontier with a maximal terminal welfare.

1.5.2 Flip-flop across time

Samuelson (1997) studied a situation where it is an intrinsic necessity to choose all of one portfolio or all of another. In contrast to commonly accepted doctrine to diversify, here a zero-one choice is required. The investor, however, can change, across time, from one stock to another. This multi-period optimization problem was formulated as an integer dynamic stochastic programming [Samuelson (1997)]. The problem is solved via the Bellman induced indirect utility functions. It is noted that, however, under the integrality constraints, a concave utility function may result in a non-concave one. But there are several interesting properties of the original utility function that is inherited by the Bellman induced indirect utility functions.

This direction has been ignored for a long time in the literature because of the wide-spread belief that diversification is the key property to reduce risk of investments. However, it is not hard to see that in reality, there are situations one may restrict one's choice. It can become an interesting topic for further investigations in future.

1.5.3 Optimal portfolios with asymptotic criteria

Merton and Samuelson (1974) proposed a model which featured a trade-off between two asymptotic criteria: the growth rate and the asymptotic variance. Konno *et al.* (1993) analyzed this trade-off by deriving the boundary of a feasible region and further designed an algorithm to compute this boundary.

However, there are two restrictions in Konno *et al.*'s model that would require further improvement in the solution in order to relax them. One is that the model assumes that securities can be modeled as geometric Brownian motion with constant parameters, which may not be true in the long-run. The other is that the model needs to continuously rebalance its portfolio which is very expensive when transaction costs are taken into consideration.

1.5.4 Multi-stage stochastic linear portfolio selection model

Dantzig and Infanger (1993) used the method of multi-stage stochastic linear program to formulate the multi-period financial asset allocation problem as follows:

$$\text{maximize } Eu(v^T)$$

$$\text{subject to } -r_i^{t-1}x_i^{t-1} + x_i^t + y_i^t - z_i^t = 0$$

$$-r_{n+1}^{t-1}x_{n+1}^{t-1} + x_{n+1}^t - \sum_{i=1}^n (1-\mu_i)y_i^t + \sum_{i=1}^n (1+v_i)z_i^t = 0$$

$$-\sum_{i=1}^{n+1} r_i^T x_i^T + v^T = 0$$

$$x1_i^t \le x_i^t \le x2_i^t,\ y1_i^t \le y_i^t \le y2_i^t,$$

$$z1_i^t \le z_i^t \le z2_i^t,\ i = 1,\cdots,n,\ t = 1,\cdots,T$$

where x_i^t is the amount of asset i in period t, y_i^t is the amount of asset i sold in period t, z_i^t is the amount of asset i bought in period t, v^T is the total value of asset at the last period T, $x1_i^t$, $x2_i^t$, $y1_i^t$, $y2_i^t$, $z1_i^t$, $z2_i^t$, $r_i^0 x_i^0$ are given parameters.

They used the techniques of the Bender decomposition and sampling to solve the above dynamic portfolio optimization and showed that their numerical results were very promising. Further investigations for efficient algorithms would make the practical use of this model for multi-stage portfolio selection.

1.5.5 Consumption-investment optimization

Some models considered both investment and consumption. Klass and Assaf (1988) discussed a multi-stage model which includes two types of assets: a riskless asset, such as bank deposit or bonds, with its growth rate r being fixed; and a risky asset with price fluctuating according to a logarithmic Brown motion with mean rate of return $b > r$. The objective is to maximize the total asset of the investor over an infinite time interval. Fleming and Zariphopoulou (1991) have further developed the above model by including consumption functions. Their objective is to maximize the total expected utility of consumption, where utility is measured by some increasing concave function $U(c)$. But they have only considered one risky asset. Akian *et al.* (1996) extended this to N numbers of risky assets. Deng, Li and Wang (2000) discussed no-arbitrage and optimal consumption-portfolio in frictional markets.

1.5.6 Universal portfolio

Cover (1991) suggested a solution (called universal portfolio) which requires no information (not even probability distribution) about the future prices of the stocks under consideration. Here a multi-stage problem is considered. The universal portfolio is regularly adjusted according to prices in previous stages. Cover considered the universal portfolio together with a class of strategies called constant re-balanced portfolio. A constant re-balanced portfolio maintains a fixed proportion of investment fund in each of the securities starting with an even proportion. An interesting result of Cover is a proof that shows, under some mild assumptions, the universal portfolio approximates the best constant re-balanced portfolio (chosen after the stock outcomes are known). He noticed that the best constant rebalanced portfolio has a better performance than any constant rebalanced portfolio, any single stock and index fund such as Down Jones Index Average (DJIA).

One weakness of Cover's algorithm is that its algorithm requires multiple dimensional integration to calculate the solution. Since the dimension grows with the number of securities, the complexity for computation is exponential. Later, Cover and Ordentlich (1996) improved the performance of their algorithm using a recurrent relation between portfolios in different stages and extended their work to utilize side information.

Helmbold and Warmuth (1998) presented an on-line investment algorithm whose time and storage requirements are a linear function of the number of securities. By this algorithm, they found that the performance of their on-line portfolio would be better than that of cover's universal portfolio. Blum and Kalai (1997) presented the performance of universal portfolios with transaction costs.

Recently, Liu, Wang and Qiu (2000a) extended and refined the results of Cover and Ordentlich (1996) and Blum and Kalai (1997) for universal portfolio with transaction costs.

1.6 Remarks and Discussions

We have considered a number of models, criteria and strategies for portfolio selection. Each model, criterion or strategy is based on certain assumptions when it is applied to portfolio selection. Some models discussed in this chapter are widely applied in portfolio choices although their assumptions may not be universally agreed upon. As the readers can see from the previous discussions, portfolio theory has been developed rapidly since Markowitz published his pioneering work on portfolio selection in 1950s. However, many problems remain to be solved. The following are what we view as most interesting and important.

(1) Transaction cost. Most often, transaction cost is either ignored or assumed to be fixed in most models [Patel and Subrahmanyam (1982), Yoshimota (1996)]. In reality, however, transaction cost is a result of several different factors such as commissions and taxes. It may also vary according to the value of a transaction. This usually results in nonlinear and non-convex function optimization problems [Li and Wang (2001), Morton and Pliska (1995), Pogue (1970)].

(2) Multi-period. Although some papers discussed above took multi-period processes of portfolio selection into consideration and made important progress, there are a lot of research topics that may help to make these methods practically more useful [Li, Li and Wang (2000a)].

(3) Incomplete information. Portfolio selection is based on uncertainty of returns of securities. Incomplete and asymmetric information is still an important aspect of further research.

(4) Influence of other factors. When an investor makes a decision for an investment, he/she has to consider a few factors, such as financing constraints, varying interest rates and the capital structure (see, e.g. [Chen, Liu, Wang and Deng (2000), Chen, Deng, Liu and Wang (2001), and Zhou, Chen and Wang (2000)]).

2. A Model for Portfolio Selection with Order of Expected Returns

2.1 Introduction

The core of the Markowitz Mean-Variance Model is to take the expected return of a portfolio as the investment return and the variance of the expected return of a portfolio as the investment risk. According to Markowitz (1952, 1959, 1987), for a given specific return level, one can derive the minimum investment risk by minimizing the variance of portfolio; or for a given risk level which the investor can tolerate, one can derive the maximum return by maximizing the expected returns of a portfolio. The main input data for the Markowitz mean-variance model are expected returns and variance of expected returns of these securities. Simplifying the number and types of the input data has been one of the main research topics in this field for the last four decades. Although some breakthroughs, such as the index model, have been implemented, all of those methods have some drawbacks due to some known reasons [Elton and Gruber (1995), Ballestero and Romero (1996)].

Transaction costs are another one of the main sources concerned by portfolio managers. Due to changes in the expectation of the future returns of securities, most of the applications of portfolio optimization involve the revision of an existing portfolio. This revision entails both purchases and sales of securities along with transaction costs. The experimental analysis indicates that ignoring the transaction costs will result in inefficient portfolios [Yoshimoto (1996)].

In this chapter, we present a new model for portfolio selection with an order of the expected returns of securities. We formulate the model in Section 2.2 and discuss how to determine the order of the expected returns of securities in Section 2.3. A genetic algorithm is designed in Section 2.4 to solve the corresponding nonconcave maximization problem. A numerical example is given to illustrate the new model and some comparisons with the results of the mean-variance model are made in Section 2.5. We give the portfolio revision process which considers the fixed transaction costs in Section 2.6. Section 2.7 gives some concluding remarks.

2.2 Model

A portfolio selection problem in the mean-variance context can be written as (P1):

$$\text{maximize} \quad (1-w)\sum_{i=1}^{n} R_i x_i - w\sum_{i=1}^{n}\sum_{j=1}^{n}\sigma_{ij}x_i x_j$$

$$\text{subject to} \quad \sum_{i=1}^{n} x_i = 1$$

$$x_i \geq 0, \ i = 1,\cdots,n$$

where n is the number of risky securities, x_i is the proportion invested in security i, R_i is the expected return of security i, σ_{ij} is the covariance of the expected return on security i and j, w is the risk aversion factor of the investor satisfying $0 \leq w \leq 1$. Obviously, the greater the factor w is, the more risk aversion the investor has. When $w=1$, the investor will be extremely conservative because in this case only the risk of his/her investment is considered and no attention is paid to the return of his/her investment. Conversely, $w=0$ means that the investor is extremely aggressive to pursue the return of his/her investment, completely ignoring the risk of investment.

The expected return and the risk of a portfolio are respectively given by

$$R_p = \sum_{i=1}^{n} R_i x_i$$

and

$$\sigma_p^2 = \sum_{i=1}^{n}\sum_{j=1}^{n}\sigma_{ij}x_i x_j$$

The input data of the maximization problem are expected returns of securities and variance-covariance matrix of the expected returns of securities in a portfolio. The following three techniques are the main methods to determine the input data.

Technique 1. According to the historical observation, put the arithmetic mean as the expected return of the security and further calculate the variance-covariance matrix of the expected returns of securities in a portfolio [Markowitz (1959)]. But for this technique, mainly two problems are to be solved:

(i) If the time horizon of the historical data of a security is very long, the influence of the earlier historical data is the same as that of the later historical data.

However, the later historical data of a security most often indicate that the performance of a corporation is more important than that of the earlier historical data.

(ii) If the historical data of a security are not much enough, one can not accurately estimate the statistical parameters due to the data scarcity.

Technique 2. Assuming that the only reason of the securities' correlation is the common response to market changes. The measure of their correlation can be obtained by relating the return on a stock to the return on a stock market index [Elton and Gruber (1995)]. The return on a stock can be broken into two components that one part is due to the market and the other is independent on the market as follows:

$$R_i = \alpha_i + \beta_i R_m + e_i$$

where α_i is the component of security $i's$ return independent of the market's performance, R_m is the rate of return on the market index, β_i is a constant that measures the expected change in R_i for a given change in R_m and e_i is the random error component. Thus, the expected return and variance of the expected return of a portfolio can be simplified as

$$R_p = \alpha_p + \beta_p R_m$$

and

$$\sigma_p^2 = \beta_p^2 \sigma_m^2 + \sum_{i=1}^n \sigma_{e_i}^2 x_i^2$$

where $\sigma_{e_i}^2$ is the variance of random error component e_i, σ_m^2 is the variance of R_m, $\alpha_p = \sum_{i=1}^n x_i \alpha_i$ and $\beta_p = \sum_{i=1}^n x_i \beta_i$. Although the single index model reduces the estimated input data from $(n^2 + n)/2$ to $3n + 2$, the fact that the correlation among securities is not only due to the single market factor has eclipsed its simplicity.

Technique 3. Using a multi-index model [Elton and Gruber (1995)] to capture some of the non-market influences such as real industries or pseudo industries. The standard form of a multi-index model can be written as

$$R_i = \alpha_i + \sum_{k=1}^m b_{ik} I_k + c_i \ , i = 1, \cdots, n$$

where α_i is the expected value of the returns not related to any index, c_i is the random component satisfying $E(c_i) = 0$, I_k is index k assumed to be pair-wise independent and b_{ik} is the measure of sensitivity to index k. Although the multi-index model can better describe historical data than the single index model, they often contain more noise than the real information when forecasting. Hence, it is not surprising that the single index model outperforms the more complicated multi-index model.

To measure the ability of different techniques to forecast the return and risk of a portfolio, Elton, Gruber and Urich (1978) compared the results of three methods as follows:

Method 1. Put historical arithmetic mean as the expected return of a security.

Method 2. Forecast expected return by estimating Beta from the prior historical period.

Method 3. Forecast expected return by estimating Beta from the prior historical period and update it via some techniques such as the techniques in Blume (1975) or Vasicek (1973) .

One of the most surprising results in Elton *et al.* (1978) is that the performance of the method based on the historical arithmetic means and their correlation matrix is the poorest. Although the single index model does not exactly describe the correlation among securities, its ability to forecast the input data outperforms the method based on the arithmetic means and their correlation matrix.

In order to improve the performance of the Markowitz mean-variance model, we consider the updating process of the arithmetic means and their correlation matrix when they are respectively put as expected returns and variance-covariance matrix of the expected returns of securities in a portfolio. Given to the past returns of a security, we can only estimate an approximate value as its expected return. If we take the expected return of a security as variable and know the order of the expected returns of securities in which the investor is interested, a new model for portfolio selection can be proposed as (P2):

$$\text{maximize} \quad (1-w)\sum_{i=1}^{n} R_i x_i - w\sum_{i=1}^{n}\sum_{j=1}^{n} \sigma_{ij} x_i x_j$$

$$\text{subject to} \quad \sum_{i=1}^{n} x_i = 1$$

$$R_i \geq R_{i+1}, \; i = 1, \cdots, n-1$$

$$a_i \leq R_i \leq b_i, \; i = 1, \cdots, n$$

$$x_i \geq 0, \; i = 1, \cdots, n$$

where $\sigma_{ij} = \frac{1}{mm}\sum_{k=1}^{mm}(R_{ik} - R_i)(R_{jk} - R_j)$ and mm is the number of the historical data of a security, R_i is the expected return of security i, R_{ik} is the $k-th$ real return of security i, and (a_i, b_i) is the range in which the expected return of security i can vary.

2.3 Order of Expected Returns

To give an order of the expected returns of securities in a portfolio and to determine the change ranges of the expected returns of securities, we will consider the following three factors:

1. Arithmetic mean. Although arithmetic means of securities should not be put as the expected returns directly, they are a good approximation.

2. Historical return tendency. If the recent historical returns of a security remain increasing, we can believe that the expected return of the security is greater than the arithmetic mean based on the historical data. However, if the recent historical returns of the security decrease as time goes, we are confident that the expected return of the security is smaller than the arithmetic mean based on the historical data.

3. Forecast of the future return of a security. The third factor influencing the expected return of a security is its estimated future return. Based on the financial report of a corporation, if we believe that the return of this corporation's stock will increase, then the expected return of this security should be larger than the arithmetic mean based on the historical data. On the contrary, if we think that the future return of this corporation's stock will decrease, the expected return of this security will be smaller than the arithmetic mean.

In order to determine the order of the expected returns of securities in a portfolio, we adopt the so-called Weight Averaging Method. First we give a weight to each of these three factors. Let h_i be the weight of factor i satisfying $h_1 + h_2 + h_3 = 1$. We have

$$R_i = h_1 * R_{ai} + h_2 * R_{ti} + h_3 * R_{fi} \, , \; i = 1, \cdots, n$$

where R_{ai} is the arithmetic mean of security i , R_{ti} is the change tendency of the return of security i , and R_{fi} is an approximation of the future expected return of security i , R_{ai} can be calculated with the historical data. Denote the arithmetic mean of the recent historical data as R_{ti}. If there is no obvious change tendency, R_{ti} can be equal to zero. The computation of derivation of R_{fi} requires some forecasts based on the financial report and individual experience. It should also be mentioned that the weights of these three factors should be given by different professionals with different experience.

With the above equation, one can give an order of the expected returns of securities in a portfolio. Meanwhile, one can also derive the low limit and the upper limit of the expected return of a security based on these three factors. If the returns of a security in the recent history have increasing tendency and also one forecasts that its future performance will be good, he/she can take R_i as a_i. This implies that the expected return of security i is larger than its arithmetic mean derived from the historical data. So the investor can put the arithmetic mean of a security as its lower limit and the maximal of the three data R_{ai}, R_{ti} and R_{fi} as its upper limit. On the other hand, if the return of a security has decreasing tendency and one forecasts that the future performance of the security is worse than the present, he/she can put the arithmetic mean of a security as its upper limit and the minimal of the three data R_{ai}, R_{ti} and R_{fi} as its lower limit.

2.4 Genetic Algorithm

Because the objective function in problem (P2) is not unimodal and the feasible set is with a particular structure, traditional optimization algorithms usually fail to find a global optimal solution of the problem. In order to avoid getting stuck at a local optimal solution, we design a genetic algorithm to find a solution to problem (P2).

Genetic algorithms are type of stochastic search methods for optimization problems based on the mechanics of natural selection and natural genetics. Genetic algorithms have demonstrated considerable success in providing good solutions to many complex optimization problems and have received more and more attention during the past three decades. For a detailed discussion, one can refer to Holland (1975), Goldberg (1989), Michalewicz (1994), and Gen and Cheng (1997).

In this section, we design a genetic algorithm to solve problem (P2). We discuss its representation structure, initialization process, evaluation function, selection process, cross and mutation operations.

Representation Structure We represent a solution by the floating point implementation in which each chromosome is coded as the following matrix of floating numbers:

$$V = \begin{pmatrix} x_1 & x_2 \cdots & x_n \\ R_1 & R_2 \cdots & R_n \end{pmatrix}$$

The chromosomes will be randomly generated in a hypercube Ω which should ensure that an optimal solution can be easily sampled from it. Usually, the generated chromosomes are not necessarily feasible solution of (P2). However, we can convert the chromosomes into feasible ones. For example, **(i)** in order to ensure that $x_1 + x_2 + \cdots + x_n = 1$, we convert the genes into new ones

$$x_i = \frac{x_i}{x_1 + x_2 + \cdots x_n}, \quad i = 1, \cdots, n$$

which make that $x_1 + x_2 + \cdots x_n = 1$; **(ii)** for $R_1 \le R_2 \le \cdots \le R_n$, we can re-arrange $\{R_1, R_2, \cdots, R_n\}$ from small to large. If the order is $R_1 \ge R_2 \ge \cdots \ge R_n$, then we re-arrange $\{R_1, R_2, \cdots, R_n\}$ from large to small.

Initialization Process. We define an integer pop_size as the number of chromosomes and initialize pop_size chromosomes randomly. It is reasonable to assume that we can predetermine a region which contains the potential optimal solution (rather than the whole feasible set). Usually, this region will be designed to have a nice shape, for example, a $2n$ dimensional hypercube, because the computer can easily sample points from a hypercube. We generate random points from the hypercube Ω pop_size times and obtain pop_size initial chromosomes $V_1, V_2, \cdots, V_{pop\_size}$ via converting them into a feasible solution of (P2).

Evaluation Function. Evaluation function, denoted by $eval(V)$, is to assign a probability of reproduction to each chromosome V so that its likelihood of being selected is proportional to its fitness relative to the other chromosomes in the population, *i.e.*, the chromosomes with higher fitness will have more chance to produce offspring by using *roulette wheel selection*. Let $V_1, V_2, \cdots, V_{pop\_size}$ be the pop_size chromosomes at the current generation. One well-known method is based on the allocation of reproductive trials according to rank rather than actual objective values. At first we calculate the objective values of the chromosomes. According to the objective values, we can re-arrange these chromosomes $V_1, V_2, \cdots, V_{pop\_size}$ from good to bad, *i.e.*, the better the chromosome is, the smaller ordinal number it has. Now let a parameter $a \in (0,1)$ in the genetic system be given, then we define the rank-based evaluation function as

$$eval(V_i) = a(1-a)^{i-1}, \quad i = 1, 2, \cdots, pop\_size$$

where $i = 1$ means the best individual and $i = pop\_size$ the worst individual.

Selection Process. The selection process is based on spinning the roulette wheel pop_size times, each time we select a single chromosome for a new population in the following way:

Step 1. Calculate the cumulative probability q_i for each chromosome V_i,

$$q_0 = 0,$$

$$q_i = \sum_{j=1}^{i} eval(V_j), \ i = 1,2,\cdots,pop\_size$$

Step 2. Generate a random real number r in $[0, q_{pop\_size}]$.

Step 3. Select the $i's$ chromosome V_i $(1 \le i \le pop\_size)$ such that $q_{i-1} < r < q_i$.

Step 4. Repeat steps 2 and 3 for pop_size times and obtain pop_size copies of chromosomes.

Crossover Operation. We define a parameter P_c of a genetic system as the probability of crossover. In order to determine the parents for crossover operation, let us do the following process repeatedly from $i = 1$ to pop_size: generating a random real number r from the interval (0,1), the chromosome V_i is selected as a parent if $r < P_c$. We denote the selected parents as $V_1, V_2, V_3 \cdots$ and divide them to the following pairs:

$$(V_1, V_2), (V_3, V_4), (V_5, V_6) \cdots$$

Let us illustrate the crossover operator on each pair by (V_1, V_2). At first, we generate a random number c from the open interval $[0,1]$, then the crossover operator on V_1 and V_2 produces two children X and Y as follows:

$$X = cV_1 + (1-c)V_2 \quad \text{and} \quad Y = (1-c)V_1 + cV_2$$

If a chromosome is not a feasible solution, then we have to revise it so that it is feasible with the procedure given in the subsection of Representation Structure.

Mutation Operation. We define a parameter P_m of a genetic system as the probability of mutation. Similar to the process of selecting parents for crossover operation, we repeat the following steps from $i = 1$ to pop_size: generating a random real number r from the interval $[0,1]$, the chromosome V_i is selected as a parent for mutation if $r < P_m$. For each selected parent, denoted by V, we mutate it by the following procedure: choose a mutation direction d in R^{2n} randomly. If $V + M$ is not in the hypercube Ω then we set M as a random number between 0 and M until it is in Ω, where M is a large positive number ensuring that the genetic operation is probabilistically complete for the chromosomes in the hypercube Ω. Here the choice of M is problem-dependent. If the above process can not yield a chromosome in the hypercube Ω in a predetermined number of iterations, then we set $M = 0$. We replace the parent V by its child $X = V + Md$. If a chromosome is not feasible, then we have to revise it to a feasible one.

Genetic Algorithm Procedure. Following selection, crossover and mutation, the new population is ready for its next evaluation. The genetic algorithm will terminate after a given number of cyclic repetitions of the above steps. We can summarize the genetic algorithm which solves the portfolio selection problem (P2) with an order constraint as follows.

Step 0. *Input parameters* pop_size, P_c *and* P_m.

Step 1. *Initialize* pop_size *chromosomes and convert them into feasible ones.*

Step 2. *Update the chromosomes by crossover and mutation operations.*

Step 3. *Calculate the objective values for all chromosomes.*

Step 4. *Compute the fitness of each chromosome by rank-based evaluation function based on the objective values.*

Step 5. *Select the chromosomes by spinning the roulette wheel.*

Step 6. *Repeat steps 2 and 5 for a given number of cycles.*

Step 7. *Report the best chromosome as the optimal solution.*

2.5 A Numerical Example

In this section, we give a numerical example to illustrate the new model proposed in this chapter. Comparisons between this new model and the mean-variance model are also presented.

Consider the following portfolio selection problem. The returns of six stocks from time *t-7* to time t are given in the following table:

Table 2.1 Security returns

Period	t-7	t-6	t-5	t-4	t-3	t-2	t-1	T
Stock 1	0.04	0.07	0.09	0.13	0.14	0.17	0.21	0.24
Stock 2	0.14	0.06	0.08	0.15	0.11	0.13	0.10	0.11
Stock 3	0.13	0.13	0.11	0.15	0.10	0.07	0.14	0.11
Stock 4	0.12	0.04	0.18	0.13	0.19	0.16	0.14	0.11
Stock 5	0.18	0.06	0.22	0.15	0.14	0.06	0.08	0.09
Stock 6	0.15	0.04	0.08	0.06	0.13	0.05	0.10	0.09

Although securities 2, 3 and 6 have not any change tendency from the above data, we can observe that securities 1 and 5 are with increasing tendencies and security 4 is with decreasing tendency. The arithmetic means of these six securities can be easily calculated. We give them in the following table:

Table 2.2 Arithmetic mean of stock returns

Stock	1	2	3	4	5	6
Arithmetic Mean	0.136	0.11	0.11	0.135	0.1225	0.0875

We can formulate the portfolio selection problem as the following maximization problem (P3):

$$\text{maximize } (1-w)\sum_{i=1}^{6} x_i R_i - w\sum_{i=1}^{6}\sum_{j=1}^{6}\sigma_{ij}x_i x_j$$

$$\text{subject to } \sum_{i=1}^{6} x_i = 1$$

$$R_i \geq R_{i+1},\ i = 1,2,\cdots,5$$

$$0.1362 \leq R_1 \leq 1$$

$$0.09 \leq R_2 \leq 0.13$$

$$0.09 \leq R_3 \leq 0.13$$

$$0 \leq R_4 \leq 0.135$$

$$0.1225 \leq R_5 \leq 1$$

$$0.05 \leq R_6 \leq 0.10$$

$$x_i \geq 0,\ i = 1,2,\cdots,6$$

To solve this problem with the genetic algorithm, we let the population size be 30, the probability of crossover P_c be 0.3 and the probability of mutation P_m be 0.2. A run of the genetic algorithm with 50000 generations obtains a solution respectively for the cases of the risk-aversion coefficient w=1, 0.8, 0.5, 0.2 and 0.01. For a comparison, we find an optimal solution by the optimization software LINGO to solve the mean-variance model (P1) in which we use the arithmetic means as the expected returns of securities. We summarize the numerical results in the following tables.

Table 2.3 Investment proportion with risk-aversion coefficient w=1

Stock	1	2	3	4	5	6	Portfolio return	Portfolio risk
Model (P1)	0	0.2598	0.4659	0.0782	0	0.2171	0.1094	0.00028
Model (P2)	0.0298	0.0441	0.4400	0.0782	0.1788	0.2291	0.1098	0.00044

Table 2.4 Investment proportion with risk-aversion coefficient w=0.8

Stock	1	2	3	4	5	6	Portfolio return	Portfolio risk
Model (P1)	0.2925	0	0	0.7075	0	0	0.1354	.00212
Model (P2)	0.2902	0	0	0.7098	0	0	0.1354	.00212

Table 2.5 Investment proportion with risk-aversion coefficient w=0.5

Stock	1	2	3	4	5	6	Portfolio return	Portfolio risk
Model (P1)	0.3943	0	0	0.6057	0	0	0.1355	0.00223
Model (P2)	0.5778	0	0	0.4222	0	0	0.1357	0.00275

Table 2.6 Investment proportion with risk-aversion coefficient w=0.2

Stock	1	2	3	4	5	6	Portfolio return	Portfolio risk
Model (P1)	0.7943	0	0	0.2057	0	0	0.1359	0.00390
Model (P2)	1	0	0	0	0	0	0.1363	0.00553

Table 2.7 Investment proportion with risk-aversion coefficient w=0.01

Stock	1	2	3	4	5	6	Portfolio return	Portfolio risk
Model (P1)	1	0	0	0	0	0	0.1363	0.00553
Model (P2)	1	0	0	0	0	0	0.1363	0.00553

It can be observed that, in the sense of investment return, the results of the new model are preferable to the results of the mean-variance model when taking the arithmetic means as the expected returns of securities although the variance in the new model is a little larger.

2.6 Portfolio Selection with Transaction Costs

Transaction cost is another important factor for an investor to take into consideration in portfolio selection. Ignoring the transaction cost in a portfolio selection model often leads to an inefficient portfolio in practice [Patel and Subrahmanyam (1982)]. If the portfolio at time t is known, we can consider its revision at time $t+1$. With the same consideration as in Markowitz (1987), we assume in the sequel that the transaction cost at time $t+1$ is a V-shaped function of the differences between the given portfolio at time t and a new portfolio at time $t+1$ as shown in Figure 2.1.

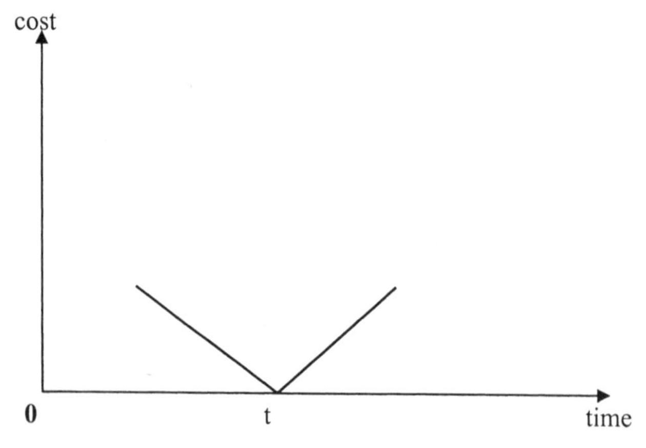

Figure 2.1 Transaction cost of securities

Thus, the transaction cost $c_{i,t+1}$ of the $i-th$ security at time $t+1$ can be expressed by

$$c_{i,t+1} = k_{i,t+1}\left|x_{i,t+1} - x_{i,t}\right|$$

where $k_{i,t+1}$ is a constant rate with respect to the $\square$ security at time t+1. So the total transaction cost of portfolio $x_{t+1} = (x_{1,t+1}, x_{2,t+1}, \cdots, x_{n,t+1})$ at time t+1 is

$$c(x_{i+1}) = \sum_{i=1}^{n} k_{i,t+1}\left|x_{i,t+1} - x_{i,t}\right|$$

For simplicity, we assume that $k_{i,t+1} = k$ for each i and t. Hence, the portfolio selection model with fixed transaction costs can be formulated as (P4):

$$\text{maximize } (1-w)\sum_{i=1}^{n} R_i x_i - w\sum_{i=1}^{n}\sum_{j=1}^{n}\sigma_{ij}x_i x_j - k\sum_{i=1}^{n}\left|x_{i,t+1} - x_{i,t}\right|$$

$$\text{subject to } \sum_{i=1}^{n} x_{i,t+1} = 1$$

$$R_{i,t+1} \geq R_{i+1,t+1},\ i = 1,\cdots,n$$

$$a_{i,t+1} \leq R_{i,t+1} \leq b_{i,t+1},\ i = 1,\cdots,n$$

$$x_{i,t+1} \geq 0,\ i = 1,\cdots,n$$

It should be mentioned that, if the above fixed transaction costs are considered, the mean-variance model (P1) can be extended as (P5):

$$\text{maximize } (1-w)\sum_{i=1}^{n} R_i x_i - w\sum_{i=1}^{n}\sum_{j=1}^{n}\sigma_{ij}x_i x_j - k\sum_{i=1}^{n}\left|x_{i,t+1} - x_{i,t}\right|$$

$$\text{subject to } x_{i,t+1} \geq 0,\ i = 1,\cdots,n$$

$$\sum_{i=1}^{n} x_{i,t+1} = 1$$

We illustrate our extended model (P4) via a numerical example and also compare the results with those derived from the mean-variance model (P5) with transaction costs. Let the historical returns be given in the following table:

Table 2.8 Security returns

Period	t-7	t-6	t-5	t-4	t-3	t-2	t-1	t	t+1
Stock 1	0.04	0.07	0.09	0.13	0.14	0.17	0.21	0.24	0.25
Stock 2	0.14	0.06	0.08	0.15	0.11	0.13	0.10	0.11	0.09
Stock 3	0.07	0.13	0.11	0.15	0.10	0.07	0.14	0.11	0.12

Stock 4	0.12	0.04	0.18	0.13	0.19	0.16	0.14	0.12	0.11
Stock 5	0.18	0.06	0.22	0.15	0.14	0.06	0.08	0.09	0.10
Stock 6	0.15	0.04	0.08	0.06	0.13	0.05	0.10	0.09	0.08

The arithmetic means of the above six securities are as follows:

Table 2.9 Arithmetic mean of stock returns

Stock	1	2	3	4	5	6
Arithmetic Mean	0.1363	0.109	0.111	0.134	0.1226	0.0873

Suppose that the transaction cost rate $k = 0.005$. With the genetic algorithm and LINGO respectively to solve (P4) and (P5), we have the following results.

Table 2.10 Investment proportion with risk-aversion coefficient w=1

Stock	1	2	3	4	5	6	Portfolio return	Portfolio risk
Model (P4)	0	0.2568	0.4639	0.0823	0	0.2178	0.1090	0.00022
Model (P5)	0.0308	0.0441	0.4100	0.0772	0.1798	0.2581	0.1091	0.00034

Table 2.11 Investment proportion with risk-aversion coefficient w=0.8

Stock	1	2	3	4	5	6	Portfolio return	Portfolio risk
Model (P4)	0.2926	0	0	0.7074	0	0	0.1371	0.00162
Model (P5)	0.2921	0	0	0.7079	0	0	0.1371	0.00162

Table 2.12 Investment proportion with risk-aversion coefficient w=0.5

Stock	1	2	3	4	5	6	Portfolio return	Portfolio risk
Model (P4)	0.3933	0	0	0.6067	0	0	0.1388	0.00177
Model (P5)	0.5768	0	0	0.4232	0	0	0.1420	0.00240

Table 2.13 Investment proportion with risk-aversion coefficient w=0.2

Stock	1	2	3	4	5	6	Portfolio return	Portfolio risk
Model (P4)	0.7943	0	0	0.2057	0	0	0.1455	.00373
Model (P5)	1	0	0	0	0	0	0.1489	.00559

Table 2.14 Investment proportion with risk-aversion coefficient w=0.01

Stock	1	2	3	4	5	6	Portfolio return	Portfolio risk
Model (P4)	1	0	0	0	0	0	0.1489	0.00559
Model (P5)	1	0	0	0	0	0	0.1489	0.00559

From the above results, we have a similar conclusion given at the end of the previous section. That is, in the sense of return, the portfolio return calculated from the new model is preferable to that derived from the mean-variance model when putting the arithmetic mean as expected return of a security.

2.7 Conclusions

A new model is proposed for portfolio selection. In the model, the input data for maximizing the return of an investment is an order of the expected returns of securities instead of the expected returns themselves. Transaction costs are also discussed. An extended model is presented for portfolio selection with transaction costs.

A genetic algorithm is designed to solve the corresponding optimization problems because these non-concave maximization problems are with a particular structure and can not be efficiently solved by the existing traditional optimization methods.

Two numerical examples are given to illustrate the new portfolio selection model and the extended one which considers transaction costs. The comparison shows that, in the sense of the portfolio return, the performance of the new models is better than that of the mean-variance models when the arithmetic means are taken as the expected returns of securities in the mean-variance models.

3. A Compromise Solution to Mutual Funds Portfolio Selection with Transaction Costs

3.1 Introduction

Since the seminal work of Markowitz (1952,1959,1987), the mean-variance methodology for the portfolio selection problem has been central to research activities of this area and has a lot of appplications during the past four decades. Extensions of the mean-variance theory include all kinds of its derivatives and the capitial asset pricing model. (see, e.g., [Xia, Wang and Deng (1998b)]).

Ballestero and Romero (1996) suggested that the application of Markowitz model to security portfolio selections normally consists of the following steps: (1) making probabilistic estimates of the future performance of securities; (2) determining the investment opportunity set as a constraint of the problem, by resorting to the mean-variance (E-V) efficient frontier; and (3) maximizing the expected utility of returns $EU(R)$ of the investor on the frontier.

In the first phase, to determine the expected return of each security, one may take the arithmetical mean of historical data, or estimate the expected return by using a regression model. For the second phase, the (E-V) efficient frontier can be formulated as a typical convex quadratic program and can be solved efficiently (in theory, by the ellipsoid method, and in reality by various practical numerical methods.) For the third phase to work, we need to make sure that there is an efficient portfolio which maximizes $EU(R)$. As pointed out in Levy and Markowitz (1979), Mean-variance model is only accurate if the returns from all securities are normally distributed or if the utility function $U(R)$ is concave and quadratic. However, as observed by Ballestero and Romero (1996), the normal distribution of returns is only a hypothesis which has not been empirically corroborated and quadratic utility functions present many logical flaws. This leaves open the possibility for alternative formulations to be applied in cases where these assumptions are not met. Ballestero and Romero, for example, suggested a compromise programming solution. They considered different measures of distance and concluded that all those compromise programming problems derive solutions corresponding to "average" investors.

The main concern that leads to this chapter is about the transaction cost, an important factor considered by investors in financial markets. Most of the cases, investors usually start with an existing portfolio and the decisions are how to re-adjust to the changes in the security market. The problem is a revision of the existing portfolio to a new one that entails both purchases and sales of securities along with transaction costs. This is a V-shaped function of the difference between the old and the new portfolios. It has been noticed in experimental analysis that ignoring transaction costs results in inefficient portfolios [Yoshimoto (1996)].

In this chapter, we consider portfolio selection with transaction costs by applying a compromise solution approach. In Section 3.2, the model is formulated and a simplified form of the objective function is discussed. We derive a compromise programming problem for portfolio selection with transaction costs in Section 3.3. In Section 3.4, we extend the compromise program to include the riskless asset which allows short sale. Section 3.5 presents some properties of the compromise soluttion. A numerical example is illustrated for deriving the comrpromise solution in Section 3.6.

3.2 Structure of Returns and Transaction Costs

As indicated in Markowitz (1952), a portfolio selection problem in a mean-variance context can be formulated as the following maximization problem (P1):

$$\begin{aligned} &\text{maximize } \textstyle\sum_{i=1}^{n} R_i x_i \\ &\text{subject to } \left(\textstyle\sum_{i=1}^{n}\sum_{j=1}^{n} S_i S_j \rho_{i,j} x_i x_j\right)^{\frac{1}{2}} \leq \gamma \qquad (3.1) \\ &\textstyle\sum_{i=1}^{n} x_i = 1 \\ &x_i \geq 0, \quad i = 1, \cdots, n \end{aligned}$$

where n is the number of risky securities, x_i is the proportion invested in security i, R_i is the expected rate of return on security i, S_i is the standard deviation of the rate of return on security i, $\rho_{i,j}$ is the correlation coefficient between the returns on securities i and j, $i, j = 1, \cdots, n$ and γ is the maximum risk the investor can accept because the standard deviation of the return characterizes the risk of a portfolio.

The expected return and the risk of a portfolio are, respectively, given by

$$R_p = \sum_{i=1}^{n} R_i x_i \tag{3.2}$$

and

$$S_p = \left(\sum_{i=1}^{n}\sum_{j=1}^{n} S_i S_j \rho_{i,j} x_i x_j\right)^{\frac{1}{2}} \tag{3.3}$$

By varying the value of γ in Problem (P1), one can find an efficient frontier which is part of a hyperbola in the $S_p - R_p$ plane, as illustrated in Figure 3.1.

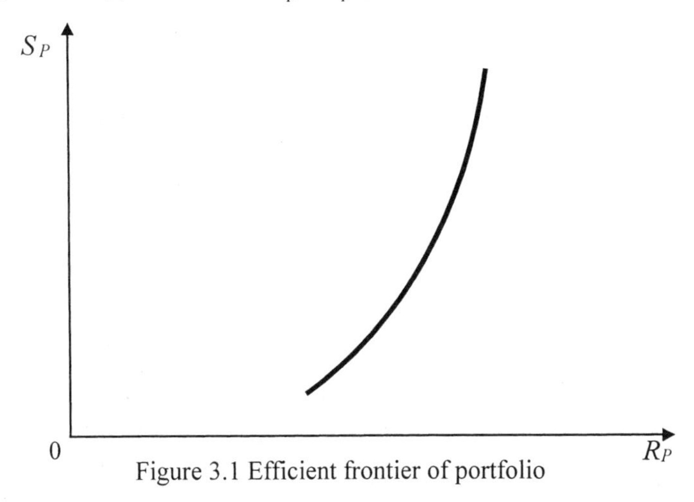

Figure 3.1 Efficient frontier of portfolio

The investor should choose his/her portfolio based on this efficient frontier. But this frontier consists of infinite number of elements and does not tell the investor which one he/she should choose from them. When the investor's utility function is known, he/she can determine a portfolio via maximizing his/her utility function on the efficient frontier. In most situations, however, the investor's utility function is not explicitly known in a portfolio selection problem. In this section, a compromise solution method is proposed to support the investor to make a decision for portfolio selection.

Elton *et al.* (1973, 1978, 1995) conducted an empirical study about the correlation coefficient structure of the returns of securities in the portfolio. They demonstrated that the performance of equal pair-wise correlation coefficients between all securities is better than the performance with the different correlation coefficients based on the historical returns. Based on this result, we assume that $\rho_{i,j} = \rho$, for all $i, j = 1, \cdots, n$ and $i \neq j$, where $0 \leq \rho \leq 1$. Thus, the variance of the expected return is

$$S_p^2 = \sum_{i=1}^{n}\sum_{j=1}^{n} S_i S_j \rho_{i,j} x_i x_j$$

$$= \rho(\textstyle\sum_{i=1}^{n}\sum_{j=1}^{n})_{i\neq j} S_i S_j x_i x_j + \sum_{i=1}^{n} S_i^2 x_i^2 \qquad (3.4)$$

$$= \rho(\textstyle\sum_{i=1}^{n} S_i x_i)^2 + (1-\rho)\sum_{i=1}^{n} S_i^2 x_i^2$$

Sharpe (1963, 1967) testified that the nonsystematic risk of the portfolio was very small compared with the systematic risk, especially when the limit of investing to each security was no more than five percent of the money which is a most common law for managing the mutual funds in many countries. He suggested that the total portfolio variability S_p should be expressed as $\beta_p \cdot S_I$ *i.e.*, the product of the portfolio Beta and the standard deviation of the market return while ignoring the nonsystematic risk $\sum_{i=1}^{n} S_i^2 x_i^2$. Because $(1-\rho)\sum_{i=1}^{n} S_i^2 x_i^2 \leq \sum_{i=1}^{n} S_i^2 x_i^2$, the standard deviation of the expected return of portfolio $x = (x_1, x_2, \cdots, x_n)$ could be approximately expressed as

$$S_p(x) = \sqrt{\rho}\textstyle\sum_{i=1}^{n} S_i x_i \qquad (3.5)$$

Transaction cost is another important factor for an investor to take into consideration in portfolio selection. Ignoring the transaction cost in a portfolio selection model often leads to an inefficient portfolio in practice. If the portfolio at time t is known, we consider its revision at time $t+1$. With the same consideration as in Markowitz (1959), we assume in this chapter that the transaction cost at time $t+1$ is a V-shaped function of the difference between the given portfolio at time t and a new portfolio at time $t+1$ as shown in Figure 3.2.

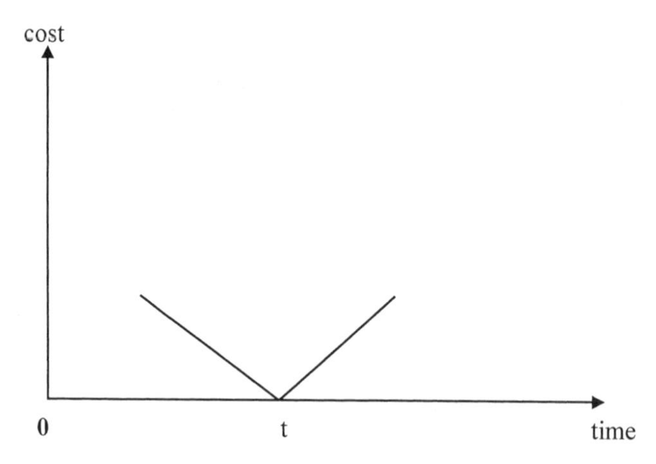

Figure 3.2 Transaction costs of securities

Hence, the transaction cost $c_{i,t+1}$ of the i -th security at time $t+1$ can be expressed by

$$c_{i,t+1} = k_{i,t+1}\left|x_{i,t+1} - x_{i,t}\right| \tag{3.6}$$

where $k_{i,t+1}$ is a constant rate with respect to the i-th security at time $t+1$. So the total transaction cost and the expected return of portfolio $x_{t+1} = (x_{1,t+1}, x_{2,t+1}, \cdots, x_{n,t+1})$ at time $t+1$ are

$$C_{t+1}(x_{t+1}) = \sum_{i=1}^{n} k_{i,t+1}\left|x_{i,t+1} - x_{i,t}\right| \tag{3.7}$$

and

$$R_p(x_{t+1}) = \sum_{i=1}^{n} R_{i,t+1}x_{i,t+1} - \sum_{i=1}^{n} k_{i,t+1}\left|x_{i,t+1} - x_{i,t}\right| \tag{3.8}$$

respectively. Based on (3.5) and (3.8), the portfolio selection problem can be reformulated as the following bi-objective nonlinear programming problem (P2):

$$\begin{aligned} &\text{maximize } R_p(x_{t+1}) = \sum_{i=1}^{n} R_{i,t+1}x_{i,t+1} - \sum_{i=1}^{n} k_{i,t+1}\left|x_{i,t+1} - x_{i,t}\right| \\ &\text{minimize } S_p(x_{t+1}) = \sqrt{\rho}\ \sum_{i=1}^{n} S_{i,t+1}x_{i,t+1} \\ &\text{subject to } \sum_{i=1}^{n} x_{i,t+1} = 1 \\ &\qquad x_{i,t+1} \le 0.05,\ i = 1, \cdots, n \\ &\qquad x_{i,t+1} \ge 0,\ i = 1, \cdots, n \end{aligned} \tag{3.9}$$

We will discuss how to solve this bi-objective programming problem in the next section.

3.3 Compromise Solution

Because (P2) is a bi-objective programming problem, we can solve this problem with several methods of multi-objective programming (MP). It is well known that a solution to a MP problem is not only dependent on the problem itself but also dependent on what a MP method is used to solve this problem. In this section, a compromise programming method is proposed to find a compromise solution to (P2). This method is easy for investors to understand because of its nice graphic explanation.

Denoting $R1_{i,t+1}$, $i=1,\cdots,20$ as the largest 20 security returns and $S1_{i,t+1}$, $i=1,\cdots,20$ as the smallest twenty security standard deviations in the stock market, then with the satisficing logic which considers trade-off between return and risk, we derive an ideal point as follows:

$$R_p^* = \frac{1}{20}\sum_{i=1}^{20} R1_{i,t+1} - \sum_{i=1}^{20} k_{i,t+1}\left|\frac{1}{20} - x_{i,t}\right| \tag{3.10}$$

$$S_p^* = \frac{\sqrt{\rho}}{20}\sum_{i=1}^{20} S1_{i,t+1}$$

i.e., using all the money investing 20 stocks with maximum returns as the ideal return of the investor while using all the money investing twenty stocks with minimum risks as the ideal risk for the investor. Correspondingly, the investor can calculate the anti-ideal point (R_{p^*}, S_{p^*}), which is the worst solution for the investor with the most averse return and risk.

We consider average risk aversion and use metric 2 to derive the following compromise programming problem (P3) :

$$\text{minimize } f = \left[\left(\frac{S_p - S_p^*}{S_{p^*} - S_p^*}\right)^2 + \left(\frac{R_p^* - R_p}{R_p^* - R_{p^*}}\right)^2 \right]^{\frac{1}{2}}$$

$$\text{subject to } \sum_{i=1}^{n} x_{i,t+1} = 1$$

$$S_p = \sqrt{\rho_t}\sum_{i=1}^{n} S_{i,t+1} x_{i,t+1}$$

$$R_p = \sum_{i=1}^{n} R_{i,t+1} x_{i,t+1} - \sum_{i=1}^{n} k_{i,t+1}\left|x_{i,t+1} - x_{i,t}\right| \tag{3.11}$$

$$x_{i,t+1} \le 0.05,\ i=1,\cdots,n$$

$$x_{i,t+1} \ge 0,\ i=1,\cdots,n$$

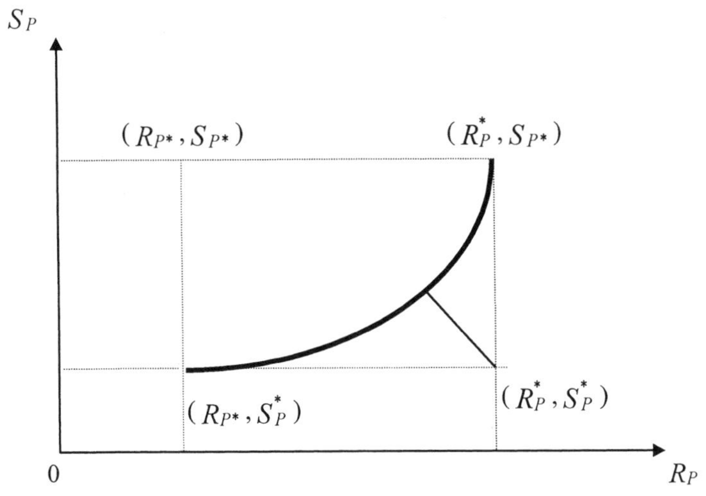

Figure 3.3 A compromise solution to portfolio selection

Let $F = f^2$ and $x_{t+1} = (x_{1,t+1}, x_{2,t+1}, \cdots, x_{n,t+1})$. Denote the feasible solution set of problem (P3) by S. It is obvious that F is a strictly increasing function of f. Thus, x_{t+1} minimizes f over S if and only if it minimizes F over S. Hence, problem (P3) is equivalent to the following minimization problem (P4):

$$\text{minimize } F(x_{t+1}) = \left(\frac{\sqrt{\rho_{t+1}} \sum_{i=1}^{n} S_{i,t+1} x_{i,t+1} - S_p^*}{S_{p^*} - S_p^*}\right)^2 + \left(\frac{R_p^* - \sum_{i=1}^{n} R_{i,t+1} x_{i,t+1} + \sum_{i=1}^{n} k_{i,t+1} |x_{i,t+1} - x_{i,t}|}{R_p^* - R_{p^*}}\right)^2$$

$$\text{subject to } \sum_{i=1}^{n} x_{i,t+1} = 1$$

$$x_{i,t+1} \le 0.05, \; i = 1, \cdots, n \tag{3.12}$$

$$x_{i,t+1} \ge 0, \; i = 1, \cdots, n$$

Consider the following minimization problem (P5):

$$\text{minimize } F(x_{t+1}) = \left(\frac{\sqrt{\rho_{t+1}}\sum_{i=1}^{n} S_{i,t+1}x_{i,t+1} - S_p^*}{S_{p^*} - S_p^*}\right)^2 + \left(\frac{R_p^* - \sum_{i=1}^{n} R_{i,t+1}x_{i,t+1} + x_{n+1,t+1}}{R_p^* - R_{p^*}}\right)^2$$

$$\begin{aligned} \text{subject to } & \sum_{i=1}^{n} k_{i,t+1}\left|x_{i,t+1} - x_{i,t}\right| \le x_{n+1,t+1} \\ & \sum_{i=1}^{n} x_{i,t+1} = 1 \\ & x_{i,t+1} \le 0.05,\, i = 1,\cdots,n \\ & x_{i,t+1} \ge 0,\ i = 1,\cdots,n \end{aligned} \tag{3.13}$$

Theorem 3.1. $(x_{1,t+1}^*,\cdots,x_{n,t+1}^*)$ is an optimal solution to (P4) if and only if there exists $x_{n+1,t+1}^*$ such that $(x_{1,t+1}^*,\cdots,x_{n,t+1}^*,x_{n+1,t+1}^*)$ is an optimal solution to (P5).

Proof. Assume that $(x_{1,t+1}^*,\cdots,x_{n,t+1}^*)$ is any optimal solution to (P4). Let $x_{n+1,t+1}^* = \sum_{i=1}^{n} k_{i,t+1}\left|x_{i,t+1} - x_{i,t}\right|$. Then $(x_{1,t+1}^*,\cdots,x_{n,t+1}^*,x_{n+1,t+1}^*)$ is a feasible solution of (P5). If it were not an optimal solution of (P5), then there would exist a feasible solution $(x_{1,t+1},\cdots,x_{n,t+1})$ of (P5) such that

$$\left(\frac{\sqrt{\rho_{t+1}}\sum_{i=1}^{n} S_{i,t+1}x_{i,t+1} - S_p^*}{S_{p^*} - S_p^*}\right)^2 + \left(\frac{R_p^* - \sum_{i=1}^{n} R_{i,t+1}x_{i,t+1} + x_{n+1,t+1}}{R_p^* - R_{p^*}}\right)^2$$

$$< \left(\frac{\sqrt{\rho_{t+1}}\sum_{i=1}^{n} S_{i,t+1}x_{i,t+1}^* - S_p^*}{S_{p^*} - S_p^*}\right)^2 + \left(\frac{R_p^* - \sum_{i=1}^{n} R_{i,t+1}x_{i,t+1}^* + x_{n+1,t+1}^*}{R_p^* - R_{p^*}}\right)^2$$

Because

$$\sum_{i=1}^{n} k_{i,t+1}\left|x_{i,t+1} - x_{i,t}\right| \le x_{n+1,t+1}$$

and

$$\sum_{i=1}^{n} k_{i,t+1}\left|x_{i,t+1}^* - x_{i,t}^*\right| = x_{n+1,t+1}^*$$

we have

$$(\frac{\sqrt{\rho_{t+1}}\sum_{i=1}^{n} S_{i,t+1}x_{i,t+1} - S_p^*}{S_{p^*} - S_p^*})^2 + (\frac{R_p^* - \sum_{i=1}^{n} R_{i,t+1}x_{i,t+1} + \sum_{i=1}^{n} k_{i,t+1}\left|x_{i,t+1} - x_{i,t}\right|}{R_p^* - R_{p^*}})^2 <$$

$$(\frac{\sqrt{\rho_{t+1}}\sum_{i=1}^{n} S_{i,t+1}x_{i,t+1}^* - S_p^*}{S_{p^*} - S_p^*})^2 + (\frac{R_p^* - \sum_{i=1}^{n} R_{i,t+1}x_{i,t+1}^* + \sum_{i=1}^{n} k_{i,t+1}\left|x_{i,t+1}^* - x_{i,t}^*\right|}{R_p^* - R_{p^*}})^2$$

which contradicts that $(x_{1,t+1}^*,\cdots,x_{n,t+1}^*)$ is optimal to (P4).

Conversely, assume that $(x_{1,t+1}^*,\cdots,x_{n,t+1}^*,x_{n+1,t+1}^*)$ is an optimal solution to (P5). It is obvious that $(x_{1,t+1}^*,\cdots,x_{n,t+1}^*)$ is feasible to (P4). If it were not an optimal solution of (P4), then there would exist a feasible solution $(x_{1,t+1},\cdots,x_{n,t+1})$ to (P4) such that

$$(\frac{\sqrt{\rho_{t+1}}\sum_{i=1}^{n} S_{i,t+1}x_{i,t+1} - S_p^*}{S_{p^*} - S_p^*})^2 + (\frac{R_p^* - \sum_{i=1}^{n} R_{i,t+1}x_{i,t+1} + \sum_{i=1}^{n} k_{i,t+1}\left|x_{i,t+1} - x_{i,t}\right|}{R_p^* - R_{p^*}})^2 <$$

$$(\frac{\sqrt{\rho_{t+1}}\sum_{i=1}^{n} S_{i,t+1}x_{i,t+1}^* - S_p^*}{S_{p^*} - S_p^*})^2 + (\frac{R_p^* - \sum_{i=1}^{n} R_{i,t+1}x_{i,t+1}^* + \sum_{i=1}^{n} k_{i,t+1}\left|x_{i,t+1}^* - x_{i,t}^*\right|}{R_p^* - R_{p^*}})^2$$

Let $x_{n+1,t+1} = \sum_{i=1}^{n} k_{i,t+1}\left|x_{i,t+1} - x_{i,t}\right|$. Then $(x_{1,t+1},\cdots,x_{n,t+1},x_{n+1,t+1})$ is a feasible solution of (P5). Since

$$\sum_{i=1}^{n} k_{i,t+1}\left|x_{i,t+1}^* - x_{i,t}^*\right| \le x_{n+1,t+1}^*$$

we get

$$(\frac{\sqrt{\rho_{t+1}}\sum_{i=1}^{n} S_{i,t+1}x_{i,t+1} - S_p^*}{S_{p^*} - S_p^*})^2 + (\frac{R_p^* - \sum_{i=1}^{n} R_{i,t+1}x_{i,t+1} + x_{n+1,t+1}}{R_p^* - R_{p^*}})^2$$

$$< (\frac{\sqrt{\rho_{t+1}}\sum_{i=1}^{n} S_{i,t+1}x_{i,t+1}^* - S_p^*}{S_{p^*} - S_p^*})^2 + (\frac{R_p^* - \sum_{i=1}^{n} R_{i,t+1}x_{i,t+1}^* + x_{n+1,t+1}^*}{R_p^* - R_{p^*}})^2$$

which contradicts that $(x_{1,t+1}^*,\cdots,x_{n,t+1}^*,x_{n+1,t+1}^*)$ is an optimal solution to (P5).

3.4 Extension of the Compromise Solution

Riskless assets such as a short-term government bill or savings account can provide the investor certain return with its standard deviation zero. If the riskless asset is considered, we give the following compromise programming problem (P6):

$$\text{minimize } F(x_{t+1}) = \left(\frac{\sqrt{\rho_{t+1}\sum_{i=1}^{n} s_{i,t+1}x_{i,t+1}} - S_p^*}{S_{p^*} - S_p^*}\right)^2 +$$

$$\left(\frac{R_p^* - \sum_{i=1}^{n} R_{i,t+1}x_{i,t+1} - x_{f,t+1}R_{f,t+1} + \sum_{i=1}^{n} k_{i,t+1}\left|x_{i,t+1} - x_{i,t}\right|}{R_p^* - R_{p^*}}\right)^2$$

$$\text{subject to } \sum_{i=1}^{n} x_{i,t+1} + x_{f,t+1} = 1 \tag{3.14}$$

$$x_{i,t+1} \le 0.05,\ i = 1,\cdots,n$$

$$x_{i,t+1} \ge 0,\ i = 1,\cdots,n$$

where $R_{f,t+1}$ and $x_{f,t+1}$ is denoted as the return and investment proportion of the riskless asset.

3.5 Properties of the Compromise Solution

Let x_{t+1}^* be an optimal solution to the compromise programming problem(P5). We summarize some properties of the compromise solution.

Property 1. Minimal Regret, i.e., x_{t+1}^* is with the minimal regret with respect to the ideal point (R_p^*, S_p^*).

This property is obvious because the value of function f at x_{t+1} measures the distance between the point $(R_p(x_{t+1}), S_p(x_{t+1}))$ and the idea point (R_p^*, S_p^*).

As mentioned before, the optimal x^*_{t+1} of (P5) is also optimal to (P4). So the point $(R_p(x^*_{t+1}), S_p(x^*_{t+1}))$ achieves the minimal distance to the ideal point (R^*_p, S^*_p) from the set $f(S) = \{f(x_{t+1}) | x_{t+1} \in S\}$. It is the minimal regret in the sense of achieving the goal (R^*_p, S^*_p).

Property 2. Pareto Efficiency, i.e., if $x_{t+1} \in S$ satisfies both $R_p(x_{t+1}) \geq R_p(x^*_{t+1})$ and $S_p(x_{t+1}) \leq S_p(x^*_{t+1})$, then $R_p(x_{t+1}) = R_p(x^*_{t+1})$ and $S_p(x_{t+1}) = S_p(x^*_{t+1})$.

The pareto efficiency implies that x^*_{t+1} is a pareto efficceient solution to (P2). This is a basic requirement for portfolio x^*_{t+1} selected by the investor.

Property 3. Scale Independence, i.e., $\bar{x}_t$ is not dependent of the unit of the expected return rates $R_{i,t}$, $i = 1,2,\cdots,n$ and the cost rate $k_{i,t+1}$, $i = 1,2,\cdots,n$.

This is obvious from the form of the objective function F in problem (P5). Actually, this form of F avoids the problem of no common measurement of two criteria $R_p(x_{t+1})$ and $S_p(x_{t+1})$ in (P2).

Property 4. No Dominance of One Criterion, i.e., both two criteria $R_p(x_{t+1})$ and $S_p(x_{t+1})$ in (P2) are considered in finding x^*_{t+1} by solving (P5).

This property is similar to the property of no dictatorship in Group Decision Making. There is no dictator of single expected return or standard deviation of the expected return of portfolio in this investment model.

3.6 A Numerical Example

To illustrate the method proposed in this chapter, we give a numerical example in this section. We use Miscrosoft EXCEL to solve the problem (P5). We discuss how an investor puts his/her money in the security market at first and then illustrate the procedure of modifying his/her portfolio when time goes on.

Suppose that at time $t = 0$, $n = 40$, $\rho_1 = 0.5$, $R_f = 0.05$, and $k_{i,0} = 0.005$, $i = 1,\cdots,40$. The expected returns and the standard deviations of the expected return of the 40 securities are respectively given as follows:

Table 3.1: Returns and risks of securities at time $t=0$

Security	Return	Risk	Security	Return	Risk	Security	Return	Risk
1	0.04	0.104	15	0.18	0.518	29	0.26	1.46
2	0.05	0.205	16	0.19	0.419	30	0.265	2.265
3	0.06	0.006	17	0.20	1.20	31	0.274	1.94
4	0.07	0.207	18	0.20	2.20	32	0.274	1.274
5	0.08	0.308	19	0.21	0.921	33	0.28	2.28
6	0.09	0.409	20	0.215	0.215	34	0.294	2.94
7	0.10	0.100	21	0.22	1.22	35	0.298	2.298
8	0.11	0.511	22	0.227	1.227	36	0.306	2.06
9	0.12	0.912	23	0.23	1.23	37	0.313	1.313
10	0.13	0.813	24	0.234	1.34	38	0.325	1.325
11	0.14	0.914	25	0.241	1.241	39	0.329	2.329
12	0.15	0.185	26	0.244	1.244	40	0.40	3.409
13	0.16	0.166	27	0.248	1.48	I^*	0.276	0.474
14	0.17	0.147	28	0.251	1.51	A^*	0.134	1.770

Note: where I^* stands for the ideal point and A^* stands for the anti-ideal point.

Without and with the riskless asset, we calculated the following results correspondingly:

Table 3.2: portfolio solutions without riskless assets at time $t=0$

Security	1	2	3	4	5	6
Proportion	0.000	0.000	0.000	0.000	0.000	0.000
Security	7	8	9	10	11	12
Proportion	0.000	0.000	0.000	0.000	0.000	0.050
Security	13	14	15	16	17	18
Proportion	0.050	0.050	0.050	0.050	0.000	0.000
Security	19	20	21	22	23	24

Proportion	0.050	0.050	0.000	0.050	0.050	0.050
Security	25	26	27	28	29	30
Proportion	0.050	0.050	0.050	0.050	0.050	0.000
Security	31	32	33	34	35	36
Proportion	0.000	0.050	0.000	0.000	0.000	0.050
Security	37	38	39	40		
Proportion	0.050	0.050	0.050	0.000		

The portfolio return from the compromise solution is 0.23785 and the risk is 1.08020.

Table 3.3: portfolio solutions with riskless assets at time $t=0$

Security	riskfree	1	2	3	4	5
Proportion	-0.150	0.000	0.000	0.050	0.000	0.000
Security	6	7	8	9	10	11
Proportion	0.000	0.050	0.000	0.000	0.000	0.000
Security	12	13	14	15	16	17
Proportion	0.050	0.050	0.050	0.050	0.050	0.050
Security	18	19	20	21	22	23
Proportion	0.000	0.050	0.050	0.050	0.050	0.050
Security	24	25	26	27	28	29
Proportion	0.050	0.050	0.050	0.050	0.050	0.050
Security	30	31	32	33	34	35
Proportion	0.000	0.000	0.050	0.000	0.000	0.000
Security	36	37	38	39	40	
Proportion	0.050	0.050	0.050	0.000	0.000	

The portfolio return from the compromise solution is 0.24252 and the risk is 1.08701.

Consider the case at time $t=1$. The expected returns and standard deviations of expected returns change as follows:

Table 3.4: Return and risks of securities at time t=1

Security	Return	Risk	Security	Return	Risk	Security	Return	Risk
1	0.35	0.114	15	0.178	0.318	29	0.259	1.546
2	0.48	0.225	16	0.191	0.419	30	0.265	2.275
3	0.056	0.080	17	0.20	1.21	31	0.271	1.84
4	0.067	0.217	18	0.205	2.21	32	0.274	1.284
5	0.078	0.208	19	0.21	0.921	33	0.28	2.48
6	0.10	0.408	20	0.218	0.210	34	0.290	2.84
7	0.11	0.11	21	0.223	1.122	35	0.294	2.278
8	0.12	0.501	22	0.227	2.237	36	0.301	2.16
9	0.132	0.911	23	0.23	1.23	37	0.309	1.310
10	0.14	0.803	24	0.232	1.44	38	0.315	1.315
11	0.145	0.914	25	0.245	1.241	39	0.322	2.309
12	0.15	0.185	26	0.248	1.224	40	0.398	3.399
13	0.156	0.167	27	0.249	1.248	I^*	0.293	0.460
14	0.17	0.147	28	0.251	1.561	A^*	0.174	1.666

Note: where I^* and A^* are the ideal point and anti-ideal point respectively.

We calculated the following results in Tables 3.5 and 3.6.

Table 3.5: portfolio solutions without riskless assets at time $t=1$

Security	1	2	3	4	5	6
Proportion	0.050	0.050	0.000	0.000	0.000	0.000
Security	7	8	9	10	11	12
Proportion	0.000	0.000	0.000	0.000	0.000	0.050
Security	13	14	15	16	17	18

Proportion	0.050	0.050	0.050	0.050	0.000	0.000
Security	19	20	21	22	23	24
Proportion	0.050	0.050	0.050	0.000	0.050	0.000
Security	25	26	27	28	29	30
Proportion	0.050	0.050	0.050	0.000	0.050	0.000
Security	31	32	33	34	35	36
Proportion	0.000	0.050	0.000	0.000	0.000	0.000
Security	37	38	39	40		
Proportion	0.050	0.050	0.050	0.050		

The portfolio return from the compromise solution is 0.25875 and the risk is 0.99670.

Table 3.6 portfolio solutions with riskless assets at time $t=1$

Security	riskfree	1	2	3	4	5
Proportion	-0.260	0.050	0.050	0.000	0.000	0.050
Security	6	7	8	9	10	11
Proportion	0.050	0.050	0.050	0.000	0.000	0.000
Security	12	13	14	15	16	17
Proportion	0.050	0.050	0.050	0.050	0.050	0.050
Security	18	19	20	21	22	23
Proportion	0.000	0.050	0.050	0.050	0.000	0.050
Security	24	25	26	27	28	29
Proportion	0.050	0.050	0.050	0.050	0.050	0.050
Security	30	31	32	33	34	35
Proportion	0.000	0.013	0.050	0.000	0.000	0.000
Security	36	37	38	39	40	
Proportion	0.000	0.050	0.050	0.000	0.000	

The portfolio return from the compromise solution is 0.26772 and the risk is 1.00749.

3.7 Conclusions

Based on some reasonable assumptions about the deviation of expected return of a portfolio and the transaction cost, we proposed in this chapter a compromise solution method to support portfolio managers for investment in security selection. A numerical example is illustrated to see how the compromise solution is derived.

4. Optimal Portfolio Selection of Assets with Transaction Costs and No Short Sales

4.1 Introduction

In the real financial markets, any movement of money between assets incurs a transaction cost proportional to the size of the transaction, paid from the bank account. In most cases, investors or portfolio managers usually start with an existing portfolio and make decisions of adjustment, probably, due to the changes of information about these securities. Since the revision entails both purchases and sales of securities that both incur transaction costs, the transaction cost is a V-shaped function of the difference between a new portfolio and the existing one. Obviously, the transaction cost has a direct impact on one's investment performance. So, transaction costs have rightly become an important factor of concern and frustration for investors. The net return of a portfolio of securities should be evaluated by taking the costs into consideration. Arnott and Wagner (1990) observed that ignoring the transaction costs might result in inefficient portfolios. The experimental analysis done by Yoshimoto (1996) also verifies this fact.

Transaction costs have also become an area of research for many authors. For example, Mao (1970b, 1970c), Jacob (1974), Brennan (1975), Levy (1978), Patel and Subrahmanyam (1982), and Morton and Pliska (1995) examined the fixed transaction costs problem. Pogue (1970), Chen, Jen, and Zionts (1971), and Yoshimoto (1996) analyzed the variable transaction costs. Recently, Dumas and Luciano (1991), Mulvey and Vladimirou (1992), Dantzig and Infanger (1993), Gennotte and Jung (1994), Akian, Menaldi, and Sulem (1995), and Atkinson and Al-Ali (1997) incorporated the transaction costs into dynamic or multi-period portfolio selection models.

On the other hand, the standard mean-variance efficient frontier of the Markowitz model has beautiful mathematical expressions and many desired properties; see, for example, Huang and Litzenberger (1988) for a complete analysis. But, in addition to including no transaction costs, the model allows short sales for assets. However, many security markets in the world such as China's Stock Markets do not allow short sales. Hence, the Markowitz model and its mean-variance approach are limited in practical applications.

The purpose of this chapter is to establish a basic theory framework for problems of portfolio selection with transaction costs and no short sales so as to enrich Markowitz theory and applications. We analyze the general model in Section 4.2 and discuss some properties of a solution in Section 4.3. An interactive method is presented in Section 4.4 and some conclusions are given in Section 4.5.

4.2. General Model Framework

We consider in this chapter a capital market with n risky assets offering random rates of returns and a risk-less asset offering a fixed rate of return. Investors (or asset managers) are assumed to be concerned with returns after taxes and transaction costs. It is assumed that there are taxes on both ordinary income and capital gains. It is also assumed that dividends and transaction costs on risky assets are paid at the end of the period and are known with certainty at the beginning of the period. An investor allocates his or her wealth among the n risky assets and the risk-less asset.

The following notations are employed:

t_g : the tax rate of marginal capital gains for the investor;

t_0 : the tax rate of the marginal ordinary income for the investor;

d_i : the dividend yield on risky asset i, equal to the monetary dividend divided by the current price;

$\tilde{r}_i$: the holding period rate of return on risky asset i, equal to the ratio of the value of the asset at the end of the period over its current value;

r_i : $E[\tilde{r}_i]$, the expected holding period rate of return on risky asset;

r_{n+1} : the holding period rate of return on the risk-less asset;

$\sigma_{ij} = \mathrm{cov}(\tilde{r}_i, \tilde{r}_j)$: the covariance between $\tilde{r}_i$ and $\tilde{r}_j$, $i, j = 1, \cdots, n$. The variance-covariance matrix $(\sigma_{ij})_{n \times n}$ is assumed to be positive semi-definite;

k_i : the constant cost per change in a proportion of the i -th risky asset, $k_i \geq 0$, $i = 1, \cdots, n$;

x_i : the proportion of the wealth the investor will invest in the i -th risky asset ($i = 1, \cdots, n$) or the risk-less asset ($i = n+1$), and

x_i^0 : the proportion of the wealth the investor already holds in the i -th risky asset ($i = 1, \cdots, n$) or the risk-less asset ($i = n+1$).

Thus, the total capital gains on portfolio $x = (x_1, \cdots, x_n, x_{n+1})$ is

$$\sum_{i=1}^{n} \tilde{r}_i x_i$$

and the total ordinary income on the portfolio is

$$\sum_{i=1}^{n} d_i x_i + r_{n+1} x_{n+1}$$

The transaction cost of the i -th risky asset c_i , as assumed by Markowitz (1987), Yoshimoto (1996), and Perold (1984), is a V-shaped function of the difference between the old portfolio $x^0 = (x_1{}^0, \cdots, x_n{}^0, x_{n+1}{}^0)$ and the new portfolio $x = (x_1, \cdots, x_n, x_{n+1})$, *i.e.*,

$$c_i = k_i \left| x_i - x_i^0 \right|, \; i = 1, \cdots, n$$

So the total transaction cost is

$$\sum_{i=1}^{n} c_i = \sum_{i=1}^{n} k_i \left| x_i - x_i^0 \right|$$

Hence, the net return after excluding the tax and the transaction cost on the portfolio is

$$(1 - t_g) \sum_{i=1}^{n} \tilde{r}_i x_i + (1 - t_0) [\sum_{i=1}^{n} d_i x_i + r_{n+1} x_{n+1}] - \sum_{i=1}^{n} k_i \left| x_i - x_i^0 \right|$$

$$= \sum_{i=1}^{n}[(1-t_g)\tilde{r}_i + (1-t_0)d_i]x_i + (1-t_0)r_{n+1}x_{n+1} - \sum_{i=1}^{n} k_i\left|x_i - x_i^0\right|$$

$$= \sum_{i=1}^{n}\tilde{R}_i x_i + R_{n+1}x_{n+1} - \sum_{i=1}^{n} k_i\left|x_i - x_i^0\right|,$$

where

$$\tilde{R}_i = (1-t_g)\tilde{r}_i + (1-t_0)d_i$$

is the after-tax rate of return on risky asset i, and

$$R_{n+1} = (1-t_0)r_{n+1}$$

is the after-tax rate of return on the risk-less asset. The expected net return after excluding the tax and the transaction cost on the portfolio is

$$g(x) = \sum_{i=1}^{n+1} R_i x_i - \sum_{i=1}^{n} k_i\left|x_i - x_i^0\right|,$$

where

$$R_i = E[\tilde{R}_i] = (1-t_g)r_i + (1-t_0)d_i$$

is the expected after-tax rate of return on risky asset i. The variance of the net return after excluding the tax and the transaction cost on the portfolio $x = (x_1, \cdots, x_{n+1})$ is thereby

$$f(x) = \sum_{i=1}^{n}\sum_{j=1}^{n} \text{cov}(\tilde{R}_i, \tilde{R}_j)x_i x_j$$

$$= (1-t_g)^2 \sum_{i=1}^{n}\sum_{j=1}^{n} \text{cov}(\tilde{r}_i, \tilde{r}_j)x_i x_j = (1-t_g)^2 \sum_{i=1}^{n}\sum_{j=1}^{n} \sigma_{ij} x_i x_j .$$

The investor expects to maximize the expected portfolio return $g(x)$ and to minimize the portfolio risk (as measured by its variance $f(x)$). Mathematically, the portfolio selection problem can be formulated as the following bi-objective programming problem (BP):

$$\text{maximize} \quad g(x) = \sum_{i=1}^{n+1} R_i x_i - \sum_{i=1}^{n} k_i\left|x_i - x_i^0\right|$$

$$\text{minimize} \quad f(x) = (1-t_g)^2 \sum_{i=1}^{n}\sum_{j=1}^{n} \sigma_{ij} x_i x_j$$

$$\text{subject to} \quad \sum_{i=1}^{n} x_i = 1$$

$$x_i \geq 0, \qquad i = 1, \cdots, n+1$$

where the first constraint implies that a fund is fully invested among risky and risk-less assets, and the other constraint is to prohibit short sales for risky assets and borrowings for risk-less asset.

Much attention has been paid to the theory and methodology of bi-objective programming or more general multi-objective programming. Here we cite the notion of efficient solutions as follows.

A portfolio $x = (x_1, \cdots, x_n, x_{n+1})$ satisfying all the constraints of (BP) is said to be feasible. A feasible portfolio x^* is said to be efficient if it is a Pareto efficient solution of (BP), that is, if there exists no other feasible portfolio x such that $g(x) \geq g(x^*)$ and $f(x) \leq f(x^*)$ with at least one strict inequality. The image of the set of all efficient portfolios under the mapping (f, g) is called the efficient frontier.

It should be pointed out that our model includes the case where there are no taxes on capital gains (or ordinary income) as a special case when setting $t_g = 0$ (or $t_0 = 0$) and the case where there are no transaction costs as a special case when setting $k_i = 0$ for $i = 1, \cdots, n$. In addition, the model also includes the case where there exists no risk-less asset as long as eliminating the variable x_{n+1} from the present model.

4.3 Solution Properties

In this section, we will establish a few properties for an efficient portfolio and the efficient frontier. These results form a basis to design an interaction method for optimal portfolio selection which will be developed in the next section.

Consider the following two single-objective optimization problems (P1) and (P2) respectively:

$$\text{minimize} \quad f(x) = (1 - t_g)^2 \sum_{i=1}^{n} \sum_{j=1}^{n} \sigma_{ij} x_i x_j$$

$$\text{subject to} \quad \sum_{i=1}^{n+1} x_i = 1$$

$$x_i \geq 0, \qquad i = 1, \cdots, n+1$$

and

$$\text{maximize} \quad g(x)=\sum_{i=1}^{n+1} R_i x_i - \sum_{i=1}^{n} k_i \left| x_i - x_i^0 \right|$$

$$\text{subject to} \quad \sum_{i=1}^{n} x_i = 1$$

$$x_i \geq 0, \quad i=1,\cdots,n+1$$

The feasible set of (P1) is the same as those of (P2) and (BP), and is denoted by X. It is clear that X is a compact convex set, $f(x)$ is a continuous convex function, and $g(x)$ is a continuous concave function. Furthermore, we consider the following two optimization problems (P_E) and (Q_V) respectively:

$$\text{minimize} \quad f(x)$$
$$\text{subject to} \quad g(x) \geq E$$
$$x \in X$$

and

$$\text{maximize} \quad g(x)$$
$$\text{subject to} \quad f(x) \leq V$$
$$x \in X$$

In an efficient security market, the more the risk of an efficient portfolio is, the more the return of the portfolio is; or, the less the return of an efficient portfolio is, the less the risk of the portfolio is. For a rational investor, certainly he or she wants to get an efficient portfolio. So, the constraints $g(x) \geq E$ in (P_E) and $f(x) \leq V$ in (Q_V) are active, that is, they become $g(x)=E$ and $f(x)=V$ at an optimal portfolios of (P_E) and (Q_V) respectively. Thus, problems (P_E) and (Q_V) are equivalent to problems $(P_E)'$ and $(Q_V)'$:

$$\text{minimize} \quad f(x)$$
$$\text{subject to} \quad g(x) = E$$
$$x \in X$$

and

$$\text{maximize} \quad g(x)$$
$$\text{subject to} \quad f(x) = V$$
$$x \in X$$

respectively. The mathematical proofs for these facts are presented below.

Let the optimal values of (P1) and (P2) be V_* and E^* respectively. Let x_* be an optimal portfolio of (Q_{V_*}), and let $E_* = g(x_*)$. Thus,

(a) $f(x_*) = V_*$;

(b) E_* is the maximal achievable value for expected return $g(x)$ without sacrificing any achievement on variance $f(x)$, while E^* is the maximal value of the expected return g at the expense of variance $f(x)$. Hence, $E_* \le E^*$. Theorem 4.3 will show that $[E_*, E^*]$ is the range of achievable values E for expected return $g(x)$ on the set of efficient portfolios.

Similarly, let x^* be an optimal portfolio of (P_{E^*}) and let $V^* = f(x^*)$. Then, $g(x^*) = E^*$, $V_* \le V^*$, and $[V_*, V^*]$ is the range of achievable values of V for variance $f(x)$ on the set of all efficient portfolios.

Theorem 4.1. (a) If $\bar{x}$ is an optimal portfolio to $(P_{\bar{E}})$ where $\bar{E} \in [E_*, E^*]$, then $g(\bar{x}) = \bar{E}$. (b) If $\bar{x}$ is an optimal portfolio to $(Q_{\bar{V}})$ where $\bar{V} \in [V^*, V_*]$, then $f(\bar{x}) = \bar{V}$.

Proof. We only prove assertion (a). The other assertion can be shown analogously. Suppose to the contrary that

$$g(\bar{x}) > \bar{E} > E_* = g(x_*). \tag{4.1}$$

Since V_* is the minimal value of (P1), we have

$$f(x_*) = V_* \le f(\bar{x}) \tag{4.2}$$

If $f(x_*) = f(\bar{x})$, then $f(\bar{x}) = V_*$, and so $\bar{x}$ is feasible to (Q_{V_*}). Thus expression (4.1) contradicts the optimality of x_* to (Q_{V_*}).

Now assume that $f(x_*) < f(\bar{x})$. By the continuity of g, (1) implies that there exists an $\hat{x} = \alpha x^* + (1-\alpha)\bar{x}$ (where $\alpha \in (0,1]$) on the closed line segment $[x^*, \bar{x}]$ which lies in the convex set X such that

$$g(\hat{x}) = \bar{E} \tag{4.3}$$

By the convexity of f and the assumption $f(x_*) < f(\bar{x})$,

$$f(\hat{x}) \le \alpha f(x_*) + (1-\alpha) f(\bar{x}) < f(\bar{x}) \tag{4.4}$$

By the concavity of g and by the expressions (4.1) and (4.3),

$$g(\frac{1}{2}\hat{x} + \frac{1}{2}\bar{x}) \ge \frac{1}{2}g(\hat{x}) + \frac{1}{2}g(\bar{x}) > \bar{E} \tag{4.5}$$

By the convexity of f and by expression (4.4),

$$f(\frac{1}{2}\hat{x} + \frac{1}{2}\bar{x}) \le \frac{1}{2}f(\hat{x}) + \frac{1}{2}f(\bar{x}) < f(\bar{x}) \tag{4.6}$$

Since $\frac{1}{2}\hat{x} + \frac{1}{2}\bar{x} \in [\hat{x}, \bar{x}] \subset [x^*, \bar{x}] \subset X$, expression (4.5) means that $\frac{1}{2}\hat{x} + \frac{1}{2}\bar{x}$ is a feasible solution to $(P_{\bar{E}})$. Hence, expression (4.6) contradicts the optimality of $\bar{x}$ to $(P_{\bar{E}})$.

Theorem 4.2. (a) The optimal portfolio set of (P_E) is identical to the one of $(P_E)'$ for each $E \in [E_*, E^*]$. (b) The optimal portfolio set of (Q_V) is identical to the one of $(Q_V)'$ for each $V \in [V_*, V^*]$.

Proof. We only show assertion (a).

First we show that each optimal portfolio of (P_E) is an optimal portfolio of $(P_E)'$. Let x^0 be optimal to (P_E). By Theorem 4.1(a), $g(x^0) = E$. So x^0 is feasible to $(P_E)'$, and hence x^0 is optimal to $(P_E)'$. We now show that every optimal portfolio of $(P_E)'$ is also an optimal portfolio of (P_E). Let x' be optimal to $(P_E)'$. Then, clearly x' is feasible to (P_E). Let x^0 be optimal to (P_E). Since we just proved that x^0 is optimal to $(P_E)'$, $f(x) = f(x^0)$. Hence, x' is an optimal solution to (P_E).

Now we can establish a necessary and sufficient condition for a portfolio to be an efficient portfolio.

Theorem 4.3. (a) A portfolio $\bar{x}$ is efficient to (BP) if and only if $\bar{x}$ is optimal to $(P_{\bar{E}})$ for some $\bar{E} \in [E_*, E^*]$. (b) A portfolio $\bar{x}$ is efficient to (BP) if and only if $\bar{x}$ is optimal to $(Q_{\bar{V}})$ for some $\bar{V} \in [V_*, V^*]$.

Proof. We only prove assertion (a).

To show the necessity, let $\bar{x}$ be an efficient solution to (BP) and let $\bar{E} = g(\bar{x})$. We claim that $g(\bar{x}) \geq g(x_*)$. Indeed, if this was not true, then $g(\bar{x}) < g(x_*)$. On the other hand, $f(\bar{x}) \geq V_* \geq f(x_*)$. Thus, x_* dominates $\bar{x}$, contradicting the efficiency of $\bar{x}$. Hence, $\bar{E} = g(\bar{x}) \geq g(x_*) = E_*$. Since it is clear that $\bar{E} \leq E^*$, we have $\bar{E} \in [E_*, E^*]$. Let $\hat{x}$ is an optimal solution to $(P_{\bar{E}})$. Thus, $f(\hat{x}) \leq f(\bar{x})$. By Theorem 4.1, $g(\hat{x}) = \bar{E} = g(\bar{x})$. Since $\bar{x}$ is efficient to (BP), it must hold that $f(\hat{x}) = f(\bar{x})$. Hence, $\bar{x}$ is an optimal solution to $(P_{\bar{E}})$.

Now we show the sufficiency. Let $\bar{x}$ be optimal to $(P_{\bar{E}})$ for some $\bar{E} \in [E_*, E^*]$. By Theorem 4.1, $g(\bar{x}) = \bar{E}$. If $\bar{x}$ was not efficient to (BP), then there would exist an $\tilde{x} \in X$ such that

$$f(\tilde{x}) \leq f(\bar{x}) \tag{4.7}$$

and

$$g(\tilde{x}) \geq g(\bar{x}) \tag{4.8}$$

with at least one strict inequality. If the strict inequality in (4.7) holds, *i.e.*, $f(\tilde{x}) < f(\bar{x})$, then $\tilde{x}$ can not be feasible to $(P_{\bar{E}})$ because $\bar{x}$ is optimal to $(P_{\bar{E}})$. Hence, $g(\tilde{x}) < \bar{E} = g(\bar{x})$, contradicting (4.8). If the strict inequality in (4.8) holds, *i.e.*, $g(\tilde{x}) > g(\bar{x}) = \bar{E}$, then $\tilde{x}$ is feasible but not optimal to $(P_{\bar{E}})$ by

Theorem 4.1. This together with the optimality of $\bar{x}$ to $(P_{\bar{E}})$ yields $f(\tilde{x}) > f(\bar{x})$ contradicting (4.7). Therefore, $\bar{x}$ is an efficient portfolio to (BP).

Theorems 4.2 and 4.3 demonstrate that an efficient portfolio can be obtained by minimizing the portfolio risk under a given level of portfolio return or by maximizing the portfolio return under a given level of portfolio risk. Consequently, the set of all efficient portfolios can be generated by parametrically solving $(P_{\bar{E}})$ (or equivalently $(P_E)'$) with E varying in $[E_*, E^*]$ or (Q_V) (or equivalently $(Q_V)'$) with V varying in $[V_*, V^*]$. In other words, the efficient portfolio frontier in variance-return space is the curve characterized by the following function:

$$V = h(E) = \min\{f(x) : g(x) \geq E, x \in X\}, \quad E \in [E_*, E^*]$$

or the following function:

$$E = H(V) = \max\{g(x) : f(x) \leq V, x \in X\}, \quad V \in [V^*, V_*].$$

Because, for an efficient portfolio, the more the risk is, the more the return is; or, the less the return is, the less the risk is, both of two above functions are strictly increasing.

Theorem 4.4. (a) The efficient frontier function $V = h(E)$ is a strictly increasing function on $[E_*, E^*]$. (b) The efficient frontier function $E = H(V)$ is a strictly increasing function on $[V_*, V^*]$.

Proof. We only prove assertion (a). Let $E_1, E_2 \in [E_*, E^*]$, and $E_1 < E_2$. Then there exist x_1 and x_2 such that

$$h(E_i) = f(x^i), \; g(x^i) = E_i, \; x_i \in X, \; i = 1,2.$$

Since $g(x^2) = E^2 > E^1$ and $x^2 \in X$,

$$x^2 \in \{x : g(x) \geq E_1, x \in X\}$$

Hence,

$$h(E_2) = f(x^2) \geq \min\{f(x) : g(x) \geq E_1, x \in X\} = h(E_1).$$

If $h(E_2) = h(E_1)$, then x^2 is optimal to (P_{E_1}). By Theorem 4.1(a), $E_2 = g(x^2) = E_1$ contradicting the assumption $E_1 < E_2$. Therefore, it must hold that $h(E_2) > h(E_1)$.

Furthermore, we show that the above two frontier functions are convex and concave respectively.

For the concepts of convexity and quasi-convexity, one can refer to, for example, Rockafellar (1970) or Bazaraa *et al.* (1993).

Theorem 4.5. (a) The efficient portfolio frontier $V = h(E)$ is a convex function on $[E_*, E^*]$. (b) The efficient portfolio frontier $E = H(V)$ is a concave function on $[V_*, V^*]$.

Proof. We only prove assertion (a). Denote

$$Z = \{(V, E) \in R^2 : x \in X, f(x) \le V, g(x) \ge E\}.$$

We claim that Z is a convex set in R^2. Let

$$(V_1, E_1),\ (V_2, E_2) \in Z,\ \ \lambda \in [0,1].$$

Thus, there exist x_1 and x_2 in X such that

$$x_i \in X\ ,\ f(x_i) \le V_i,\ g(x_i) \ge E_i,\ \ i = 1,2\ .$$

Since X is a convex set,

$$\lambda x_1 + (1-\lambda) x_2 \in X\ .$$

Since $f(x)$ is a convex function,

$$f(\lambda x_1 + (1-\lambda)x_2) \le \lambda f(x_1) + (1-\lambda) f(x_2)\ \le \lambda V_1 + (1-\lambda) V_2$$

Since $g(x)$ is a concave function,

$$g(\lambda x_1 + (1-\lambda)x_2) \ge \lambda g(x_1) + (1-\lambda) g(x_2)\ \ge \lambda E_1 + (1-\lambda) E_2$$

Hence,

$$\lambda(V_1, E_1) + (1-\lambda)(V_2, E_2)\ = (\lambda V_1 + (1-\lambda)V_2, \lambda E_1 + (1-\lambda)E_2) \in Z\ .$$

That is, Z is convex. Hence, by Theorem 5.3 of Rockafellar (1970),

$$\varphi(E) = \inf\{V : (V, E) \in Z\}$$

is a convex function on R^1. It is clear that

$$\varphi(E) = h(E)\text{, for any } E \in [E_*, E^*]$$

Hence, $h(E)$ is a convex function on the interval $[E_*, E^*]$.

Among all the efficient portfolios, the investor will choose one as his or her best investment strategy that maximizes his or her utility; that is, he or she solves the following problem (P3):

$$\text{maximize} \quad U[f(x), g(x)]$$
$$\text{subject to} \quad x \in S$$

where S is the set of all efficient portfolios and U is the investor's preference function or utility function originally defined on the payoff set

$$Y = \{(V, E) : V = f(x), E = g(x), x \in X\}$$

such that for any two portfolios x^1 and x^2, $U[f(x^1), g(x^1)] > U[f(x^2), g(x^2)]$, if the investor prefers x^1 to x^2. It is easy to see that such a U is strictly increasing with respect to return g and strictly decreasing with respect to risk *f*. It is not hard to verify that (P3) is equivalent to the problem (P4):

$$\text{maximize} \quad U[f(x), g(x)]$$
$$\text{subject to} \quad x \in X$$

that is, maximizing the utility over S is identical to that over X. The maximal utility is

$$\max_{x \in S} U[f(x), g(x)] = \max_{E \in [E_*, E^*]} U[h(E), E] = \max_{V \in [V_*, V^*]} U[V, H(V)]$$

by Theorem 4.3 and the definitions of h and H. Define two new functions as follows:

$$e(E) = U[h(E), E], \quad E \in [E_*, E^*]$$

and

$$v(V) = U[V, H(V)], \quad V \in [V_*, V^*]$$

Recalling the convex set Z in the proof of Theorem 4.5, obviously $Y \subset Z$.

Theorem 4.6. Let the utility function $U(f, g)$ be defined over the convex set Z, decreasing with respect to *f* and increasing with respect to g. If U is (strictly) concave [(strongly) quasi-concave respectively] on *Z*, then (a) function $e(E)$ is (strictly) concave [or (strongly) quasi-concave respectively] on $[E_*, E^*]$; (b) function $v(V)$ is (strictly) concave [or (strongly) quasi-concave respectively] on $[V_*, V^*]$

Proof. We only show the case that $e(E)$ is concave. Let $E_1, E_2 \in [E_*, E^*]$ and $\lambda \in [0,1]$. Since h is convex by Theorem 4.5,

$$h(\lambda E_1 + (1-\lambda)E_2) \le \lambda h(E_1) + (1-\lambda)h(E_2) \tag{4.9}$$

Noting that $(h(E_1), E_1)$, $(h(E_2), E_2) \in Y \subset Z$, by the convexity of Z, we have

$$(\lambda h(E_1) + (1-\lambda)h(E_2), \lambda E_1 + (1-\lambda)E_2) \in Z$$

Thus, by inequality (4.9) and the decreasingness of U with respect to its first variable,

$$\begin{aligned} &U[h(\lambda E_1 + (1-\lambda)E_2), \lambda E_1 + (1-\lambda)E_2] \\ &\ge U[\lambda h(E_1) + (1-\lambda)h(E_2), \lambda E_1 + (1-\lambda)E_2] \end{aligned} \tag{4.10}$$

and by the concavity of U,

$$\begin{aligned} &U[\lambda h(E_1) + (1-\lambda)h(E_2), \lambda E_1 + (1-\lambda)E_2] \\ &\ge \lambda U[h(E_1), E_1] + (1-\lambda)U[h(E_2), E_2] \end{aligned} \tag{4.11}$$

Hence, by the definition of $e(E)$ and inequalities (4.10) and (4.11),

$$\begin{aligned} e(\lambda E_1 + (1-\lambda)E_2) &= U[h(\lambda E_1 + (1-\lambda)E_2), \lambda E_1 + (1-\lambda)E_2] \\ &\ge U[\lambda h(E_1) + (1-\lambda)h(E_2), \lambda E_1 + (1-\lambda)E_2] \\ &\ge \lambda U[h(E_1), E_1] + (1-\lambda)U[h(E_2), E_2] \\ &= \lambda e(E_1) + (1-\lambda)e(E_2) \end{aligned}$$

Therefore, $e(E)$ is concave.

Theorem 4.6 indicates that the (strict) risk averse investor's utility $e(E)$ (or $v(V)$) on the efficient frontier is a (strictly) concave function of return E (or risk V) over the interval $[E_*, E^*]$ (or $[V_*, V^*]$) because (strict) risk aversion implies a (strict) concave utility function U [see, e.g., Huang (1988)].

4.4 An Interactive Method

In this section, we develop an efficient interactive method for optimal portfolio selection. For brevity, we proceed with our analysis using only the function $v(V)$ defined in Section 4.2. Interested readers can also develop a method using the function $e(E)$ in a similar way.

According to the results established in the previous section, the optimal portfolio selection problem is reduced to determine the maximum point of $v(V)$ where V belongs to the interval $[V_*, V^*]$. Unfortunately, the exact analytic form of

$v(V)$ is not known because the investor's utility function U is not known explicitly. However, the optimal portfolio selection problem can still be solved by using a search technique which requires only paired preference comparison rather than the exact function values.

To develop such a method, we need the following result.

Theorem 4.7. For the (strictly) concave function $v(V)$ defined on the finite interval $[V_*, V^*]$, let V_1 and V_2 be two values in the interval such that $V_1 < V_2$. Then, (a) if $v(V_1) < v(V_2)$, the maximum of $v(V)$ will not lie in the interval $[V_*, V_1]$; (b) if $v(V_1) > v(V_2)$, the maximum of $v(V)$ will not lie in the interval $[V_2, V^*]$.

Proof. If the conclusion of (a) was not true, then the maximum of $v(V)$, say $\bar{V}$, would belong to $[V_*, V_1]$. Thus $V_1 \in [\bar{V}, V_2]$ and hence, there would exist $\alpha \in [0,1]$ such that $V_1 = \alpha\bar{V} + (1-\alpha)V_2$. By the (strict) concavity of $v(V)$,

$$v(V_1) \geq \alpha v(\bar{V}) + (1-\alpha)v(V_2) \geq \alpha v(V_2) + (1-\alpha)v(V_2) = v(V_2)$$

which contradicts the assumption that $v(V_1) < v(V_2)$.

The proof of (b) is analogous.

Denote by $\succ$ as "prefer to", which is a preference relation represented by the investor's utility function U.

Choose any two points V_1 and V_2 in the interval $[V_*, V^*]$ such that $V_1 < V_2$. Let $y^{(i)} = (V_i, H(V_i)), \quad i = 1,2$. If $y^{(2)} \succ y^{(1)}$, then

$$v(V_1) = U[y^{(1)}] < U[y^{(2)}] = v(V_2)$$

Hence, by Theorem 4.7, the maximum of $v(V)$ can not lie in the interval $[V_*, V_1]$. Eliminating the region $[V_*, V_1]$, the search interval, where the maximum lies, reduces to the interval (V_1, V^*). Similarly, if $y^{(1)} \succ y^{(2)}$, the search interval becomes (V_*, V_2). This provides a convenient way of ruling out a portion of the efficient frontier. All that are needed is an interaction with the investor whether

$$y^{(2)} \succ y^{(1)} \text{ or } y^{(1)} \succ y^{(2)} \text{ or } y^{(1)} \sim \text{(indifferent to) } y^{(2)}$$

Using his/her response, we can narrow down the search interval. Repeating this process, we can more and more exactly estimate the location of the maximum. Certainly, the effect depends on the choice of paired points. There have been many methods of searching points [see, e.g., Bazaraa (1993)]. Among them, the Fibonacci section method and the Golden section method are very popular and efficient. If the Golden section method is used to generate the two new points in (V_1, V^*), one of the new points will turn to be V_2. With the Golden section method, we will be able to narrow down the search interval to less than 10% of the original interval with just 5 iterations; in 10 stages, we will be within 1%; etc. Thus, for a specified level of the final search interval, we have a finite, rapidly converging procedure using only paired comparison of implicit preference from the investor. See Figure 4.1 for an illustration of the method.

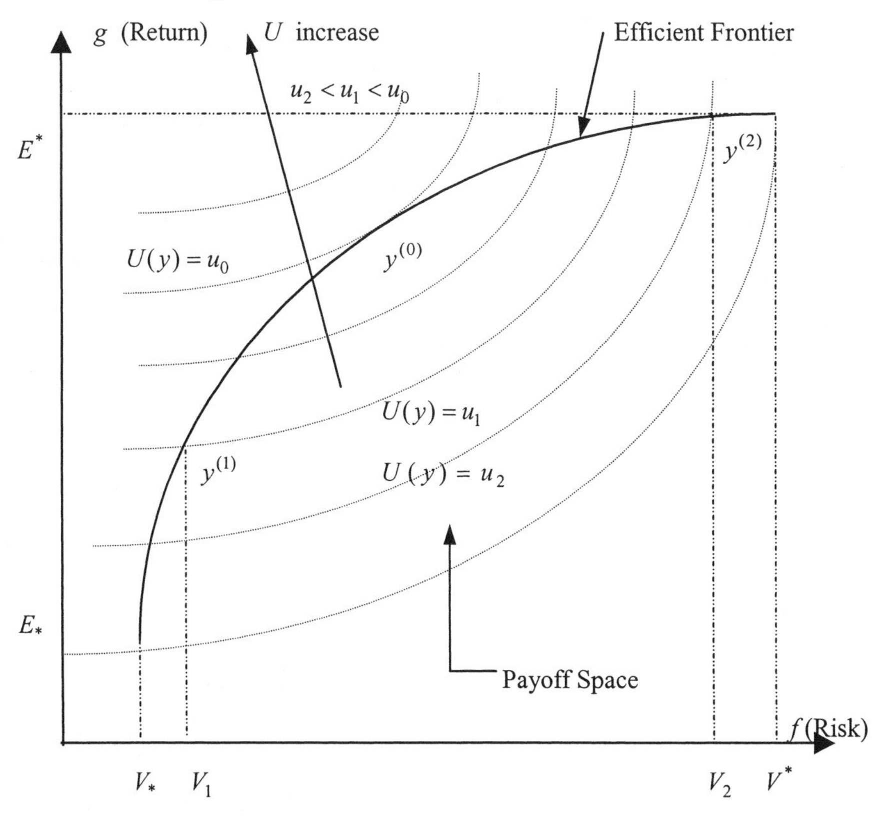

Figure 4.1. Interactive method

Let $\delta = \frac{\sqrt{5}-1}{2} \approx 0.618$ (Golden section ratio). We state the procedure of the algorithm using the Golden section method as follows:

Step 1. Solve (P1) *and set* $V_* =$ *the optimal value of* (P1). *Solve* (P2) *and set* $E^* =$ *the optimal value of* (P2).

Step 2. Solve (P_{E^*}) *and set* $V_* =$ *the optimal value of* (P_{E^*}).

Step 3. Set $a_1 = V_*$ *and* $b_1 = V^*$. *Given the tolerance error* $\varepsilon > 0$.

Step 4. Calculate $\lambda_1 = \alpha_1 + (1-\delta)(b_1 - a_1)$ *and* $\mu_1 = \alpha_1 + \delta(b_1 - \alpha_1)$. *Solve* (Q_V) *for* $V = \lambda_1$ *and set* $H(\lambda_1) =$ *its optimal value. Solve* (Q_V) *for* $V = \mu_1$ *and set* $H(\mu_1) =$ *its optimal value. Set* $y^{(1)} = (\lambda_1, H(\lambda_1))$ *and* $z^{(1)} = (\mu_1, H(\mu_1))$. *Set* $k = 1$.

Step 5. Ask the investor to specify whether $y^{(k)} \succ z^{(k)}$ *or* $z^{(k)} \succ y^{(k)}$ *or* $y^{(k)} \sim z^{(k)}$.

Step 6. If $y^{(k)} \succ z^{(k)}$ *or* $y^{(k)} \sim z^{(k)}$, *go to Step 7. If* $z^{(k)} \succ y^{(k)}$, *go to Step 8.*

Step 7. If $\mu_k - \alpha_k \le \varepsilon$, *stop and produce output of maximum* λ_k. *Otherwise, set* $\alpha_{k+1} = \alpha_k$, $b_{k+1} = \mu_k$, $\mu_{k+1} = \lambda_k$, $z^{(k+1)} = y^{(k)}$. *Calculate* $\lambda_{k+1} = \alpha_{k+1} + (1-\delta)(b_{k+1} - \alpha_{k+1})$. *Solve* (Q_V) *for* $V = \lambda_{k+1}$ *and set* $H(\lambda_{k+1}) =$ *its optimal value. Set* $y^{(k+1)} = (\lambda_{k+1}, H(\lambda_{k+1}))$ *and go to Step 9.*

Step 8. If $b_k - \lambda_k \le \varepsilon$, *stop and produce output of maximum* μ_k. *Otherwise, set* $\alpha_{k+1} = \lambda_k$, $b_{k+1} = b_k$, $\lambda_{k+1} = \mu_k$, $y^{(k+1)} = z^{(k)}$. *Calculate* $\mu_{k+1} = \alpha_{k+1} + \delta(b_{k+1} - \alpha_{k+1})$. *Solve* (Q_V) *for* $V = \mu_{k+1}$ *and set* $H(\mu_{k+1}) =$ *its optimal value. Set* $z^{(k+1)} = (\mu_{k+1}, H(\mu_{k+1}))$ *and go to Step 9.*

Step 9. Set $k = k+1$; *go back to Step 5.*

We give a numerical example to illustrate the method presented above.

Example 4.1. Let $n=4$, $t_g = 0$, $k_1 = k_2 = k_3 = k_4 = 0.005$, $R_1 = 1.13$, $R_2 = 1.035$, $R_3 = 1.14$, $R_4 = 1.,175$, $R_5 = 1.10$, $x_1^0 = x_2^0 = x_3^0 = x_4^0 = 0$, $\sigma_{11} = 0.20$, $\sigma_{22} = 0.15$, $\sigma_{33} = 0.23$, $\sigma_{44} = 0.52$, $\sigma_{12} = \sigma_{21} = 0.10$, $\sigma_{13} = \sigma_{31} = 0.15$, $\sigma_{14} = \sigma_{41} = 0.01$, $\sigma_{23} = \sigma_{32} = -0.16$, $\sigma_{24} = \sigma_{42} = -0.20$, $\sigma_{34} = \sigma_{43} = 0.10$

Then,

$$f(x) = 0.20x_1^2 + 0.15x_2^2 + 0.23x_3^2 + 0.52x_4^2 \\ +2(0.10x_1x_2 + 0.15x_1x_3 + 0.01x_1x_4 - 0.16x_2x_3 - 0.20x_2x_4 + 0.10x_3x_4)$$

$$g(x) = 1.13x_1 + 1.035x_2 + 1.14x_3 + 1.175x_4 + 1.10x_5 - 0.005(x_1 + x_2 + x_3 + x_4),$$

$$X = \{x = (x_1, x_2, x_3, x_4, x_5) : x_1 + x_2 + x_3 + x_4 + x_5 = 1, x_i \geq 0, i = 1, \cdots, 5\}$$

In order to simulate the interaction process, we assume that the investor's implicit utility function is the strictly concave function $U(f, g) = g - f^2$ which is strictly decreasing with respect to f and strictly increasing with respect to g.

The optimal portfolio (*i.e.* the solution of (P4)) is

$$\bar{x} = (x_1, x_2, x_3, x_4, x_5) = (0.24219, 0.000, 0.10810, 0.34110, 0.30861)$$

with payoff $\bar{y} = (f(\bar{x}), g(\bar{x})) = (0.09180, 1.133715)$ and utility $U(\bar{y}) = 1.125288$.

The optimal portfolio using the new method is computed as follows:

Step 1. Solving (P1) and (P2) gives $V_* = 0$ and $E^* = 0.17$.

Step 2. Solving (P_{E^*}) yields $V^* = 0.52$

Step 3. Set $\alpha_1 = V_* = 0$ and $b_1 = V^* = 0.52$. Assume that the search interval is to be reduced to 10%, *i.e.*, $e = 0.052$

Step 4. Calculate $\lambda_1 = \alpha_1 + (1 - \delta)(b_1 - \alpha_1) = 0.19864$ and

$$\mu_1 = \alpha_1 + \delta(b_1 - \alpha_1) = 0.32136$$

Solving (Q_V) for $V = \lambda_1$ and $V = \mu_1$, we get $H(\lambda_1) = 1.14957$ and $H(\mu_1) = 1.15983$, $y^{(1)} = (\lambda_1, H(\lambda_1)) = (0.199864, 1.14957)$, $z^{(1)} = (\mu_1, H(\mu_1))$ $= (0.32136, 1.15983)$.

Step 5. Ask the investor to compare $y^{(1)}$ with $z^{(1)}$ according to his or her implicit preference function which is assumed to be $U(f, g) = g - f^2$.

Step 6.

Since $U(y^{(1)}) = 1.11011$ and $U(z^{(1)}) = 1.05656$, $y^{(1)} \succ z^{(1)}$

Step 7. Since $\mu_1 - \alpha_1 \geq \varepsilon$, iteration has to be continued.

⋮

The results are summarized in Table 4.1. When *k*=5, we get search interval [λ_5, b_5]=[0.07588,0.12276] and an approximate optimal solution μ_5 =0.09378 which is an approximate value of $f(\bar{x})$.

Table 4.1. Summary of computations

k	α_k	b_k	λ_k	μ_k	$y^{(k)}$
1	0	0.52	0.19864	0.32136	(0.19864,1.14957)
2	0	0.32136	0.12276	0.19864	(0.12276,1.13899)
3	0	0.19864	0.07588	0.12276	(0.07588,10.13065)
4	0	0.12276	0.04489	0.07588	(0.04689,1.12410)
5	0.04689	0.12276	0.07588	0.09378	(0.07588,1.13065)

$z^{(k)}$	$U(y^{(k)})$	$U(z^{(k)})$	%
(0.32136,1.15983)	1.11011	1.05656	61.8
(0.19864,1.14957)	1.12392	1.11011	38.2
(0.12276,1.13899)	1.12489	1.12392	23.6
(0.07588,1.13065)	1.12190	1.12489	14.6
(0.09378,1.13408)	1.12489	1.12529	9.0

Here k stands the k-th iteration and % the remaining interval of the original.

4.5. Conclusions

In this chapter, we have extended the portfolio optimization setting of the mean-variance approach proposed by Markowitz to the case with transaction costs, in a general setting, as a V-shape function of the differences between the old and new portfolios, and with prohibiting short sales for assets. A number of properties of an efficient portfolio and the efficient frontier are established. Based on these results, an interaction method which requires only pair-wise comparison of preference from the investor is presented. A numerical example for illustrating the method is given by using the procedure of the algorithm.

5. Portfolio Frontier with Different Interest Rates for Borrowing and Lending

5.1 Introduction

The theory of portfolio selection of Markowitz (1952) applies mean and variance to characterize return and risk for a combination of more than two financial assets traded in a frictionless economy. Unlimited short selling is allowed and the rates of return on these assets are assumed to have finite variances. Rothschild and Stiglitz (1970, 1971) introduced the concept of second degree stochastic dominance. They showed that when there are more than two risky assets, if there exists a portfolio of assets that the second degree stochastically dominates all the portfolio with the same expected rate of return, then this dominant portfolio must have the minimum variance among all the portfolios. This observation is one of the motivations for characterizing portfolios that have the minimum variance for various levels of expected rate of return. The study of the mean-variance efficiency by Gonzalez-Gaverra (1973), Merton (1972) and Roll (1977) expanded Markowitz model to discuss various issues in portfolio management. To understand the formulation of mean-variance as return-risk, Chamberlain (1983) made an effort to characterize the complete family of probability distributions that are necessary and sufficient for the expected utility of terminal wealth to be a function only of the mean and variance of terminal wealth or for mean-variance utility functions. Epstein (1985) showed that the mean-variance utility functions were implied by a set of decreasing absolute risk aversion postulates.

Huang and Litzenberger (1988) stated the theory of portfolio selection in the case that there exists a riskless asset rate r, then the investor might short-sell his/her riskless asset, invest the risky assets, and obtain minimum variance. This portfolio is in the efficient portfolio frontier. If the expected rate of return $E[\tilde{r}_p]$ of the portfolio p is strictly higher than the riskless rate r, then the investor will short-sell his/her riskless asset, invest the risky assets, and obtain minimum variance; this portfolio is in the efficient portfolio frontier. If the expected rate of return $E[\tilde{r}_p]$ of the portfolio p is equal to the riskless rate r, then the investor will invest the riskless asset and the variance of this portfolio is zero. If the expected rate of return $E[\tilde{r}_p]$ of the portfolio p is strictly lower than the riskless rate r, then this portfolio is not in the efficient portfolio frontier.

The original result of Markowitz was derived in a discrete time, frictionless economy with the same interest rates for borrowing and lending. In reality, investors may be charged a higher interest rate for borrowing money than the interest rate for saving money. Even though many publications assumed the same riskless interest rate for borrowing/lending, the discrepancy between borrowing and lending is crucial for the operations of financial institutions. This chapter extends the portfolio study of Markowitz to the case of different interest rates for borrowing and lending.

The results of this chapter, which is different from Huang and Litzenberger (1988), stated the theory of portfolio selection in the case of different interest rates for borrowing and lending as follows. If the expected rate of return $E[\tilde{r}_p]$ of the portfolio p is strictly higher than the riskless rate of lending r_l, then the investor will short-sell his riskless asset, invest the risky assets, and obtain minimum variance; this portfolio is in the efficient portfolio frontier which is a continuous (smooth) curve with various simple and basic portfolio frontiers in four cases. If the expected rate of return $E[\tilde{r}_p]$ of the portfolio p is equal to the riskless rate of lending r_f, then the investors will invest the riskless asset; the variance of this portfolio is zero. If the expected rate of return $E[\tilde{r}_p]$ of the portfolio p is strictly lower than the riskless rate of lending r_f, then this portfolio is not in the efficient portfolio frontier which is a continuous curve with various simple and basic portfolio frontiers in four cases.

In Section 5.2, we introduce a few notations and definitions. In Section 5.3, we examine the problem with different interest rates for borrowing and lending where the riskless borrowing rate is higher than the riskless lending rate. The quadratic program for establishing the portfolio frontier is solved in the closed form by the Kuhn-Tucker optimality condition. Section 5.4 concludes the chapter with a few remarks and discussions.

5.2. Preliminaries

We follow the notations in Huang and Litzenberger (1988). In general, we consider $N+1$ assets: N risky assets and one riskless asset with different interest rates for borrowing and lending. Unlimited short selling is allowed and that the rates of return on these assets have finite variances and unequal expectation. In this section, we first review the case without the riskless asset to introduce some notations and properties useful to our discussions in the sequel of this chapter; then, we review the case of the same interest rate for the riskelss asset.

5.2.1 Portfolio frontier without riskless asset

The random rate of return on the n-th risky asset is $\tilde{r}_n$ for any $n \in \mathrm{N} = \{1, \cdots, N\}$.

Its expected rate and variance are $E[\tilde{r}_n]$ and $\sigma^2(\tilde{r}_n)$, respectively. Let $\tilde{r}$ denote the N-vector of rates of return on the N risky assets, e denote the N-vector of expected rates of return on the N risky assets and V the variance-covariance matrix:

$$\tilde{r} = \begin{pmatrix} \tilde{r}_1 \\ \vdots \\ \tilde{r}_N \end{pmatrix} \quad e = \begin{pmatrix} E[\tilde{r}_1] \\ \vdots \\ E[\tilde{r}_N] \end{pmatrix} \quad V = \begin{pmatrix} \mathrm{cov}(\tilde{r}_1, \tilde{r}_1) & \cdots & \mathrm{cov}(\tilde{r}_1, \tilde{r}_N) \\ \vdots & & \vdots \\ \mathrm{cov}(\tilde{r}_N, \tilde{r}_1) & \cdots & \mathrm{cov}(\tilde{r}_N, \tilde{r}_N) \end{pmatrix}$$

Let I be an N-vector of all ones. $e = E[\tilde{r}] \neq \alpha \mathrm{I}$ for any $\alpha \in R^1$ and $V = \mathrm{cov}(\tilde{r}, \tilde{r})$. It is also assumed that the rate of return on any asset as a random variable can not be expressed as a linear combination of the rates of return on the other assets. Under this assumption, the asset returns are linearly independent and their variance-covariance matrix V is nonsingular. The matrix V is thus positive definite (in general it is positive semi-definite). The rate of return on portfolio p is $\tilde{r}_p = \sum_{n \in \mathrm{N}} w_n \tilde{r}_n = w_p^T \tilde{r}$. The expected rate of return on portfolio p is

$$E[\tilde{r}_p] = \sum_{n \in \mathrm{N}} w_n E[\tilde{r}_n] = w_p^T e$$

and its variance is

$$\sigma^2(\tilde{r}_p) = \sigma^2(w_p^T \tilde{r}) = \mathrm{cov}(w_p^T \tilde{r}, w_p^T \tilde{r}) = w_p^T \mathrm{cov}(\tilde{r}, \tilde{r}) w_p^T = w_p^T V w_p .$$

A frontier portfolio has the minimum variance among the portfolios that have the same expected rate of return. Thus, w_p, the N-vector portfolio weights of the efficient portfolio p, is the solution to the following quadratic programming problem:

$$\text{minimize} \quad \frac{1}{2} w^T V w$$

$$\text{subject to} \quad w^T e = E[\tilde{r}_p]$$

$$w^T \mathrm{I} = 1$$

Here we minimize the portfolio variance subject to the constraints that the portfolio expected rate of return is equal to $E[\tilde{r}_p]$ and that the portfolio weights sum to unity. Short sales (*i.e.*, negative portfolio weights) are permitted. Its unique solution of the quadratic programming problem is

$$w_p = \frac{CE[\tilde{r}_p] - A}{D} V^{-1} e + \frac{B - AE[\tilde{r}_p]}{D} V^{-1} \mathrm{I}$$

where

$$A = e^T V^{-1} \mathrm{I} = \mathrm{I}^T V^{-1} e$$

$$B = eV^{-1} e$$

$$C = \mathrm{I} V^{-1} \mathrm{I}$$

$$D = BC - A^2 .$$

The variance of the efficient portfolio p is

$$\sigma^2(\tilde{r}_p) = \frac{C}{D}(E[\tilde{r}_p] - \frac{A}{C})^2 + \frac{1}{C}$$

That is,

$$\frac{\sigma^2(\tilde{r}_p)}{\frac{1}{C}} - \frac{(E[\tilde{r}_p] - \frac{A}{C})^2}{\frac{D}{C^2}} = 1$$

which is a hyperbola with center $(0, \frac{A}{C})$ and asymptotes $E[\tilde{r}_p] = \frac{A}{C} \pm \sqrt{\frac{D}{C}} \sigma(\tilde{r}_p)$, in the space of standard deviation and expected rate of return. This portfolio frontier in the space of mean and standard deviation is shown in Figure 5.1.

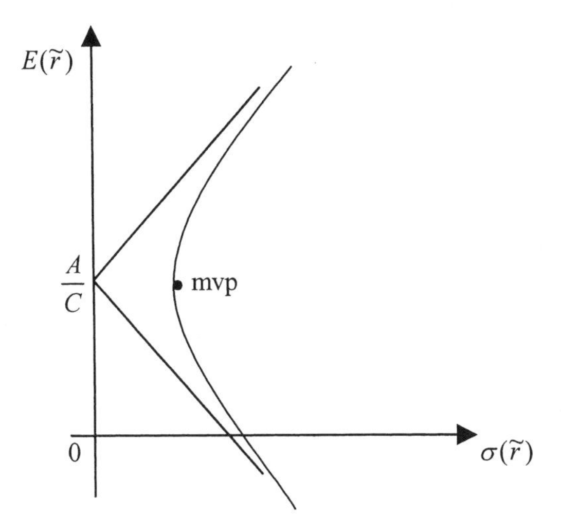

Figure 5.1 Portfolio frontier in the $\sigma(\tilde{r}) - E[\tilde{r}]$ space

5.2.2 General model

Let r_b and r_l be the riskless borrowing rate and the riskless lending rate of return on the riskless asset, respectively. Assume that the riskless borrowing rate is higher than the riskless lending rate: $r_b \geq r_l$. Let w_p denote the N-vector portfolio weights of p on the risky assets. Then $1 - w_p^T \mathrm{I}$ is the weight of the riskless asset. And the rate of return on portfolio p is

$$\tilde{r}_p = \sum_{n \in \mathrm{N}} w_n \tilde{r}_n + (1 - \sum_{n \in \mathrm{N}} w_n) r = w_p^T \tilde{r} + (1 - w^T \mathrm{I}) r$$

where

$$r = \begin{cases} r_l & if \quad 1 - w^T \mathrm{I} \geq 0 \\ r_b & if \quad 1 - w^T \mathrm{I} < 0 \end{cases}$$

If $1 - w^T \mathrm{I} \geq 0$ $(w^T \mathrm{I} \leq 1)$, the investor short selles the portfolio of N risky assets and invests (lends) the proceeds in the riskless asset. Hence, $r = r_l$. If $1 - w^T < 0$ $(w^T \mathrm{I} > 1)$, the investor long selles the portfolio of N risky assets and short selles (borrows) the proceeds in the riskless asset. So $r = r_b$.

Therefore, the expected rate of return on portfolio p is

$$E[\tilde{r}_p] = \sum_{n\in N} w_n E[\tilde{r}_n] + (1 - \sum_{n\in N} w_n) r$$

$$= w_p^T e + (1 - w^T \mathrm{I}) r$$

It follows that $w_p^T (e - r\mathrm{I}) = E[\tilde{r}_p] - r$. The variance of portfolio p is

$$\sigma^2(\tilde{r}_p) = \sigma^2(w_p^T \tilde{r} + (1 - w^T \mathrm{I}) r) = \sigma^2(w_p^T \tilde{r}) = \mathrm{cov}\left(w_p^T \tilde{r}, w_p^T \tilde{r}\right)$$

$$= w_p^T \mathrm{cov}(\tilde{r}, \tilde{r}) w_p = w_p^T V w_p$$

A portfolio p has the minimum variance among the portfolios that have the same expected rate of return if and only if w_p, the N-vector portfolio weights of p, is the solution to the quadratic programming problem:

$$\text{minimize } \frac{1}{2} w^T V w$$

$$\text{subject to } w^T e + (1 - w^T \mathrm{I}) r = E[\tilde{r}_p]$$

5.2.3 Case of the same interest rate

Let the riskless borrowing rate be equal to the riskless lending rate, that is, let $r = r_b = r_l$. The unique set of the portfolio weights for the portfolio p having an expected rate of return of $E[\tilde{r}_p]$ is

$$w_p = \frac{E[\tilde{r}_p] - r}{H} V^{-1}(e - r\mathrm{I})$$

where $H = (e - r\mathrm{I})^T V^{-1} (e - r\mathrm{I}) = B - 2Ar + Cr^2 > 0$. It follows that the variance of the rate of return on portfolio p is $\sigma^2(\tilde{r}_p) = \dfrac{(E[\tilde{r}_p] - r)^2}{H}$. Therefore,

$$\sigma(\tilde{r}_p) = \begin{cases} -\dfrac{E[\tilde{r}_p] - r}{\sqrt{H}}, & \mathit{if} \quad E[\tilde{r}_p] \le r \\ \dfrac{E[\tilde{r}_p] - r}{\sqrt{H}}, & \mathit{if} \quad r \le E[\tilde{r}_p] \end{cases}$$

The portfolio frontier of all the assets is composed of two half-lines emanating from the point $(0,r)$ in the $\sigma(\tilde{r})-E[\tilde{r}]$ plane with slopes $\sqrt{H}$ and $-\sqrt{H}$, respectively.

(α) $r>\frac{A}{C}$. This case is presented graphically in Figure 5.2, where e' is the tangent point of the half line $r-\sqrt{H}\sigma(\tilde{r}_p)$ and the portfolio frontier of all the risky assets where $E[\tilde{r}_e]=\frac{Ar-B}{Cr-A}=\frac{A}{C}-\frac{\frac{D}{C^2}}{r-\frac{A}{C}}$. Any portfolio on the half-line $r-\sqrt{H}\sigma(\tilde{r}_p)$ involves a long position in portfolio e'. Any portfolio on the line $r+\sqrt{H}\sigma(\tilde{r}_p)$ involves short-selling portfolio e' and investing the proceeds in the riskless asset.

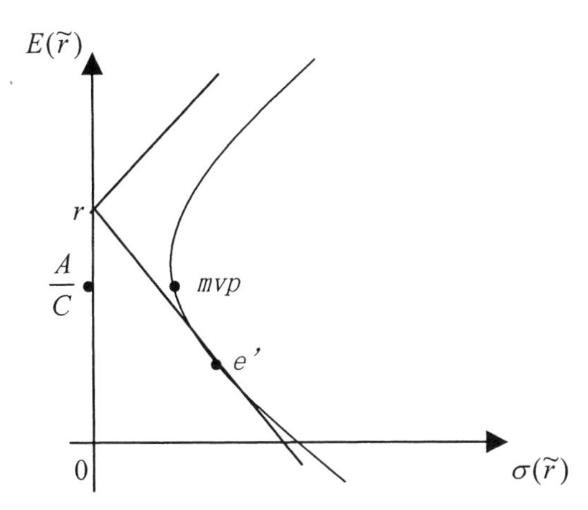

Figure 5.2 Portfolio frontier when $r>\frac{A}{C}$.

(β) $r=\frac{A}{C}$. Then $H=B-2Ar+Cr^2=B-2A\frac{A}{C}+C\frac{A^2}{C^2}=\frac{D}{C}>0$ and $E[\tilde{r}_p]=\frac{A}{C}\pm\sqrt{\frac{D}{C}}\sigma(\tilde{r}_p)$ are the asymptotes of the portfolio frontier of the risky assets which is the portfolio frontier of all the assets graphed in Figure 5.3. Any portfolio on the portfolio frontier of all the assets involves investing everything in the riskless asset and holding an arbitrage portfolio of the risky assets - a portfolio whose weights sum to zero.

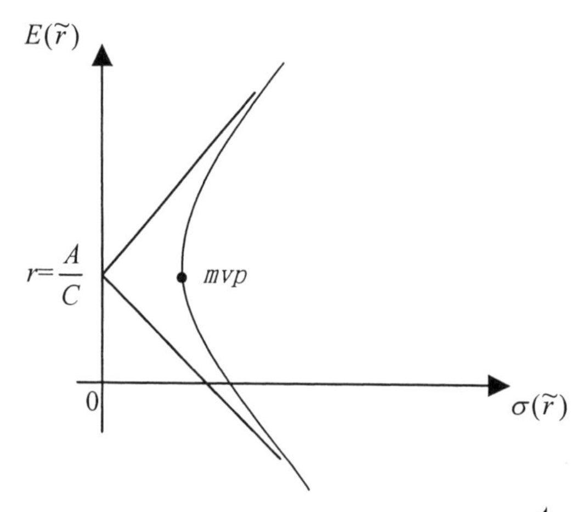

Figure 5.3 Portfolio frontier when $r = \frac{A}{C}$

(γ) $r < \frac{A}{C}$. This case is presented graphically in Figure 5.4, where e is the tangent point of the half line $r + \sqrt{H}\sigma(\tilde{r}_p)$ and the portfolio frontier of all the risky assets where $E[\tilde{r}_e] = \frac{Ar - B}{Cr - A} = \frac{A}{C} - \frac{\frac{D}{C^2}}{r - \frac{A}{C}}$. Any portfolio on the half line $r - \sqrt{H}\sigma(\tilde{r}_p)$ involves short-selling portfolio e and investing the proceeds in the riskless asset. Any portfolio on the half line $r + \sqrt{H}\sigma(\tilde{r}_p)$ other than those on the line segment $\overline{re}$ involves short-selling the riskless asset and investing the proceeds in portfolio e.

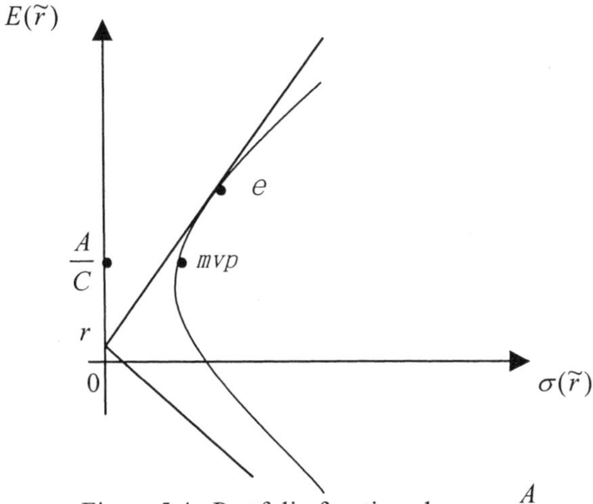

Figure 5.4 Portfolio frontier when $r < \frac{A}{C}$

5.3. Portfolio Frontier with Different Interest Rates

When the riskless borrowing rate is strictly higher than the riskless lending rate, that is, $r_b > r_l$, the quadratic program (QP) is as follows:

$$\text{minimize } \frac{1}{2} w^T V w$$

$$\text{subject to } w^T e + (1 - w^T \mathrm{I}) r = E[\tilde{r}_p]$$

where

$$r = \begin{cases} r_l & \text{if } \quad 1 - w^T \mathrm{I} \geq 0 \\ r_b & \text{if } \quad 1 - w^T \mathrm{I} < 0 \end{cases}.$$

In the quadratic program (QP) , when $w^T \mathrm{I} \leq 1 r = r_l$. And when $w^T \mathrm{I} > 1$, $r = r_b$. This quadratic program (QP) can be solved by the quadratic programs (QP1) and (QP2) respectively:

$$\text{minimize } \frac{1}{2} w^T V w$$

$$\text{subject to } \quad w^T e + (1 - w^T \mathrm{I}) r_l = E[\tilde{r}_p]$$

and

$$\text{minimize} \quad \frac{1}{2} w^T V w$$

$$\text{subject to} \quad w^T e + (1 - w^T \mathrm{I}) r_b = E[\tilde{r}_p]$$

$$w^T \mathrm{I} > 1$$

Of the solutions to the above two quadratic programs, the one with smaller variance is the solution to the quadratic program (QP). We will discuss (QP1) and (QP2) separately.

5.3.1 Solution to (QP1)

For the quadratic program (QP1), $\nabla\{1 - w^T \mathrm{I}\} = -I$ and $\nabla\{w^T e + (1 - w^T \mathrm{I}) r_l - E[\tilde{r}_p]\} = e - r_l I$. Since $e \neq \alpha \mathrm{I}$ for any $\alpha \in R$, rank $\{e - r_l \mathrm{I}, -\mathrm{I}\} = 2$. Thus, any w in the constraint of the quadratic program (QP1) is a regular point and we can apply the Kuhn-Tucker optimality condition to find a unique solution of (QP1).

Forming the Lagrangian function, w_p is the solution to the following unconstrained minimization problem:

$$\text{minimize} \quad L - \frac{1}{2} w^T V w \quad \lambda\{w^T e + (1 - w^T \mathrm{I}) r_l - E[\tilde{r}_p]\} - \mu(1 - w^T \mathrm{I})$$

where $\mu \geq 0$. By the Kuhn-Tucker optimality condition, we get

$$\frac{\partial L}{\partial w} = V w^p - \lambda(e - r_l \mathrm{I}) + \mu \mathrm{I} = 0 \tag{5.1}$$

$$\frac{\partial L}{\partial \lambda} = -\{w_p^T e + (1 - w_p^T \mathrm{I}) r_l - E[\tilde{r}_p]\} = 0 \tag{5.2}$$

$$1 - w_p^T \mathrm{I} \geq 0 \tag{5.3}$$

$$\mu \geq 0 \tag{5.4}$$

$$\mu(1 - w_p^T \mathrm{I}) = 0 \tag{5.5}$$

Solving Equation (5.1), we have

$$w_p = V^{-1}[\lambda(e - r_l \mathrm{I}) - \mu \mathrm{I}] \tag{5.6}$$

That is, $w_p^T = [\lambda(e - r_l \mathrm{I})^T - \mu \mathrm{I}^T]V^{-1}$. Right-multiplying both sides by $e - r_l \mathrm{I}$ and combining it with Equation (5.2), we have

$$\lambda(e - r_l \mathrm{I})^T V^{-1}(e - r_l \mathrm{I}) - \mu \mathrm{I}^T V^{-1}(e - r_l \mathrm{I}) = E[\tilde{r}_p] - r_l$$

Let $H_l = (e - r_l \mathrm{I})^T V^{-1}(e - r_l \mathrm{I}) = B - 2Ar_l + Cr_l^2 > 0$. By the relation (5.2), this reduces to

$$\lambda H_l - \mu(A - r_l C) = E[\tilde{r}_p] - r_l \tag{5.7}$$

We consider three cases where $A - r_l C$ (<0, =0, >0) according to $\mu = 0$ and $\mu > 0$.

First, when $\mu = 0$, the conditions (5.4) and (5.5) are satisfied. From Relation (5.7), $\lambda = \dfrac{E[\tilde{r}_p] - r_l}{H_l}$. Substituting λ and $\mu = 0$ into Equality (5.6), we obtain the unique solution of portfolio weights w_p for the portfolio p having an expected rate of return of $E[\tilde{r}_p]$:

$$w_p = \frac{E[\tilde{r}_p] - r_l}{H_l} V^{-1}(e - r_l \mathrm{I}) \tag{5.8}$$

Hence, the variance of the rate of return on portfolio p is

$$\sigma^2(\tilde{r}_p) = w_p^T V w_p = \frac{(E[\tilde{r}_p] - r_l)^2}{H_l} \tag{5.9}$$

Note that

$$w_p^T \mathrm{I} = \frac{E[\tilde{r}_p] - r_l}{H_l}(e - r_l \mathrm{I})^T V^{-1} \mathrm{I} = \frac{E[\tilde{r}_p] - r_l}{H_l}(A - r_l C).$$

By (1.3), we have $(E[\tilde{r}_p] - r_l)(A - r_l C) \le H_l$. Therefore,

$$E[\tilde{r}_p](A - r_l C) \le B - r_l A \tag{5.10}$$

(1.a.1) If $A - r_l C < 0$, then $r_l > \dfrac{A}{C}$ and $E[\tilde{r}_p] \ge \dfrac{B - r_l A}{A - r_l C}$.

$$\sigma(\tilde{r_p}) = \begin{cases} -\dfrac{E[\tilde{r_p}]-r_l}{\sqrt{H_l}} & \text{if} \quad \dfrac{B-r_l A}{A-r_l C} \le E[\tilde{r_p}] \le r_l \\ \dfrac{E[\tilde{r_p}]-r_l}{\sqrt{H_l}} & \text{if} \quad r_l \le E[\tilde{r_p}] \end{cases}$$

Thus, the portfolio frontier of all the assets is composed of a closed line segment and a half-line in the $\sigma(\tilde{r_p}) - E[\tilde{r_p}]$ plane: the closed line segment between the point e_l' and the point $(0, r_l)$ involving the proceeds in the risky assets and the riskless asset without borrowing and lending; and the half-line emanating from the point $(0, r_l)$ with slope $\sqrt{H_l}$ involving short-selling portfolio e_l' and investing (lending) the proceeds in the riskless asset.

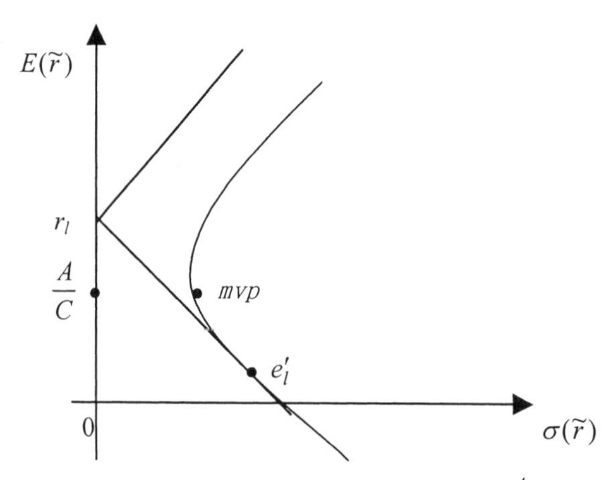

Figure 5.5 Portfolio frontier when $r_l > \dfrac{A}{C}$

(1.b.1) If $A - r_l C = 0$, then $r_l = \dfrac{A}{C}$ and Inequality (5.10) holds for any $E[\tilde{r_p}]$.

$$H_l = B - 2Ar_l + Cr_l^2 = B - 2A\frac{A}{C} + C\frac{A^2}{C^2} = \frac{D}{C} > 0 .$$

$$\sigma(\tilde{r}_p) = \begin{cases} -\dfrac{E[\tilde{r}_p] - \dfrac{A}{C}}{\sqrt{\dfrac{D}{C}}}, & if \quad E[\tilde{r}_p] \le \dfrac{A}{C} \\ \dfrac{E[\tilde{r}_p] - \dfrac{A}{C}}{\sqrt{\dfrac{D}{C}}}, & if \quad \dfrac{A}{C} \le E[\tilde{r}_p] \end{cases}$$

The portfolio frontier of all the assets is the asymptote of the portfolio frontier of the risky assets graphed in Figure 5.6. Any portfolio on the portfolio frontier of all the assets involves investing everything in the riskless asset and holding an arbitrage portfolio of the risky assets - a portfolio with total weight sum equal to zero.

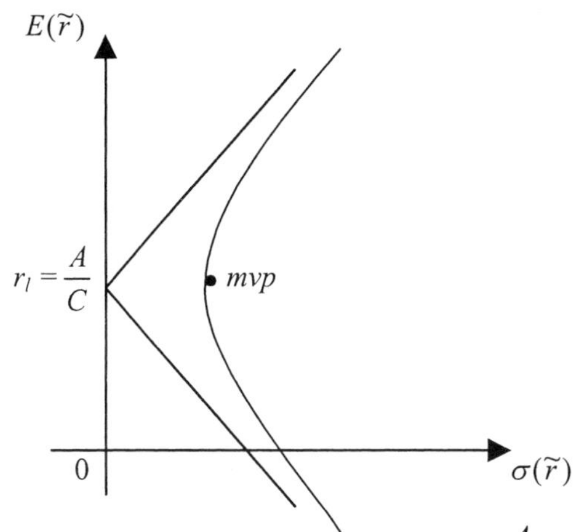

Figure 5.6 Portfolio frontier when $r_l = \dfrac{A}{C}$

(1.c.1) If $A - r_l C > 0$, then $r_l < \dfrac{A}{C}$ and $E[\tilde{r}_p] \le \dfrac{B - r_l A}{A - r_l C}$.

$$\sigma(\tilde{r}_p) = \begin{cases} -\dfrac{E[\tilde{r}_p] - r_l}{\sqrt{H_l}}, & if \quad E[\tilde{r}_p] \le r_l \\ \dfrac{E[\tilde{r}_p] - r_l}{\sqrt{H_l}}, & if \quad r_l \le E[\tilde{r}_p] \le \dfrac{B - r_l A}{A - r_l C} \end{cases}$$

Then portfolio frontier of all assets is composed of an half-line and a closed line segment in the $\sigma(\tilde{r}_p) - E[\tilde{r}_p]$ plane: the half-line emanating from the point $(0, r_l)$

with slope $-\sqrt{H_l}$ involving short-selling portfolio e_l and investing the proceeds in the riskless asset; and the closed line segment between the point $(0, r_l)$ and the point e_l involving the proceeds in the risky assets and the riskless asset without borrowing and lending.

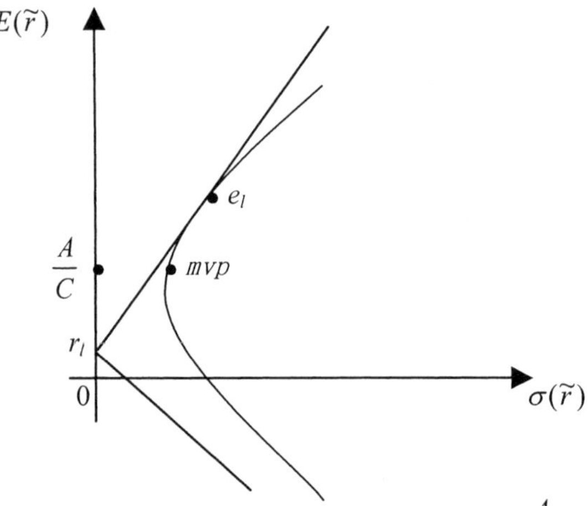

Figure 5.7 Portfolio frontier when $r_l < \frac{A}{C}$

Secondly, when $\mu > 0$, Inequality (5.4) is satisfied. From Equation (5.5), $w_p^T \mathrm{I} = 1$. Inequality (5.3) and Equation (5.5) are satisfied. Hence,

$$\lambda(e - r_l\ \mathrm{I})^T V^{-1}\mathrm{I} - \mu \mathrm{I}^T V^{-1}\mathrm{I} = 1$$

By (5.6), we have

$$\lambda(A - r_l C) - \mu C = 1 \tag{5.11}$$

Solving λ and μ from the system of Equations (5.7) and (5.11), we have

$$\lambda = \frac{(E[\tilde{r}_p] - r_l)C - (A - r_l C)}{D} = \frac{E[\tilde{r}_p]C - A}{D}$$

$$\mu = \frac{(E[\tilde{r}_p] - r_l)(A - r_l C) - H_l}{D} = \frac{E[\tilde{r}_p](A - r_l C) - (B - r_l A)}{D}$$

Substituting λ and μ into Equation (5.6) gives the unique set of portfolio weights for the portfolio p having an expected rate of return of $E[\tilde{r}_p]$:

$$w_p = \frac{1}{D} V^{-1}[(E[\tilde{r}_p]C - A)e + (B - E[\tilde{r}_p]A)\mathrm{I})] \tag{5.12}$$

Therefore, the variance of the rate of return on portfolio p is

$$\sigma^2(\tilde{r}_p) = w_p^T V w_p = \frac{C}{D}(E[\tilde{r}_p] - \frac{A}{C})^2 + \frac{1}{C} \tag{5.13}$$

$$\sigma(\tilde{r}_p) = \sqrt{\frac{C}{D}(E[\tilde{r}_p] - \frac{A}{C})^2 + \frac{1}{C}}$$

Note that $\mu > 0$ is the same as $E[\tilde{r}_p](A - r_l C) - (B - r_l A) > 0$. That is,

$$E[\tilde{r}_p](A - r_l C) > B - r_l A \tag{5.14}$$

(1.a.2) If $A - r_l C < 0$, then $r_l > \frac{A}{C}$ and $E[\tilde{r}_p] < \frac{B - r_l A}{A - r_l C}$. The portfolio frontier of all the assets is composed of a part below the point e_l' of the right-branch curve of the hyperbola, in the space of standard deviation and expected rate of return, with center $(0, \frac{A}{C})$ and asymptotes $E[\tilde{r}_p] = \frac{A}{C} \pm \sqrt{\frac{D}{C}}\sigma(\tilde{r}_p)$ involving short-selling the riskless asset and investing the proceeds in the portfolio of the risky assets. The portfolio frontier in the space of mean and standard deviation is presented in Figure 5.8.

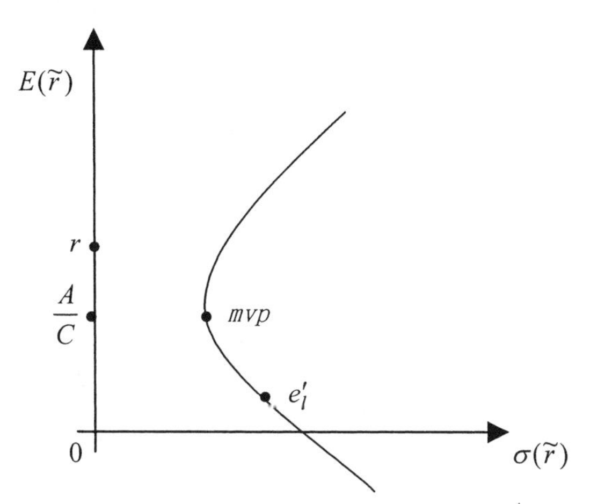

Figure 5.8 Portfolio frontier when $r_l > \frac{A}{C}$

(1.b.2) If $A - r_l C = 0$, then $r_l = \frac{A}{C}$ and Inequality (5.14) does not hold.

(1.c.2) If $A - r_l C > 0$, then $r_l < \frac{A}{C}$ and $E[\tilde{r}_p] > \frac{B - r_l A}{A - r_l C}$. The portfolio frontier of all the assets is composed of a part above the point e_l of the right-branch curve of the hyperbola, in the space of standard deviation and expected rate of return, with center $(0, \frac{A}{C})$ and asymptotes $E[\tilde{r}_p] = \frac{A}{C} \pm \sqrt{\frac{D}{C}} \sigma(\tilde{r}_p)$ involving short-selling (lending) the riskless asset and investing the proceeds in the portfolio of the risky assets. The portfolio frontier in the space of mean and standard deviation is presented in Figure 5.9.

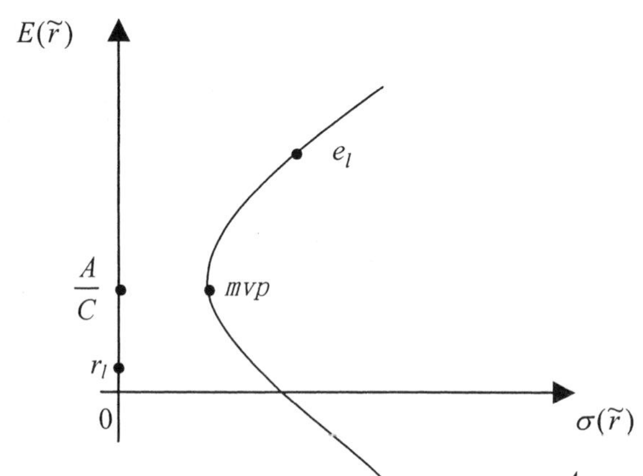

Figure 5.9 Portfolio frontier when $r_l < \frac{A}{C}$

We can summarize the above discussion for the cases $\mu = 0$ and $\mu > 0$.

(1.a) If $A - r_l C < 0$, then $r_l > \frac{A}{C}$. The unique set of portfolio weights for the portfolio p having an expected rate of return of $E[\tilde{r}_p]$ is

$$w_p = \begin{cases} \frac{1}{D} V^{-1}[(E[\tilde{r}_p]C - A)e + (B - E[\tilde{r}_p]A)\mathrm{I})], & if \quad E[\tilde{r}_p] < \frac{B - r_l A}{A - r_l C} \\ \frac{E[\tilde{r}_p] - r_l}{H_l} V^{-1}(e - r_l \mathrm{I}), & if \quad \frac{B - r_l A}{A - r_l C} \leq E[\tilde{r}_p] \end{cases}$$

and the standard deviation of the rate of return on portfolio p is

$$\sigma(\tilde{r}_p) = \begin{cases} \sqrt{\frac{C}{D}(E[\tilde{r}_p] - \frac{A}{C})^2 + \frac{1}{C}}, & if \quad E[\tilde{r}_p] < \frac{B - r_l A}{A - r_l C} \\ -\frac{E[\tilde{r}_p] - r_l}{\sqrt{H_l}}, & if \quad \frac{B - r_l A}{A - r_l C} \le E[\tilde{r}_p] \le r_l \\ \frac{E[\tilde{r}_p] - r_l}{\sqrt{H_l}}, & if \quad r_l \le E[\tilde{r}_p] \end{cases}$$

The portfolio frontier of all the assets is composed of a curve, a closed line segment and a half-line in the $\sigma(\tilde{r}_p) - E[\tilde{r}_p]$ plane: the curve part below the point e'_l of the right-branch curve of the hyperbola with center $(0, \frac{A}{C})$ and asymptotes $E[\tilde{r}_p] = \frac{A}{C} \pm \sqrt{\frac{D}{C}}\sigma(\tilde{r}_p)$ involving short-selling the riskless asset and investing the proceeds in portfolio of the risky assets; the closed line segment between the point e'_l and the point $(0, r_l)$ involving the proceeds in the risky assets and the riskless asset without borrowing and lending; and the half-line emanating from the point $(0, r_l)$ with slope $\sqrt{H_l}$ involving short-selling portfolio e'_l and investing (lending) the proceeds in the riskless asset.

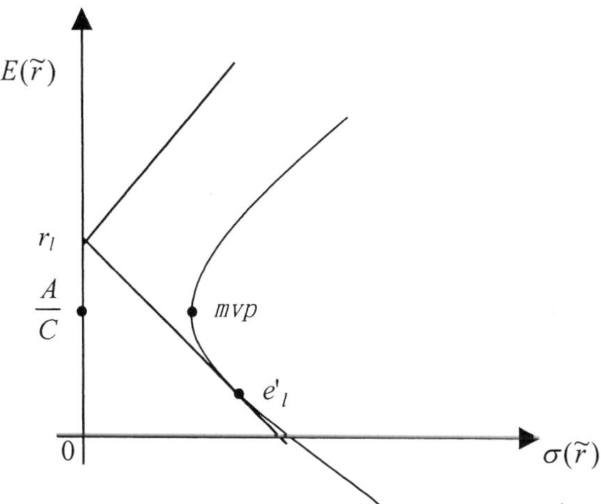

Figure 5.10 Portfolio frontier when $r_l > \frac{A}{C}$

(1.b) If $r - r_l C = 0$, then $r_l = \frac{A}{C}$. The unique set of portfolio weights for the portfolio p having an expected rate of return of $E[\tilde{r}_p]$ is as follows:

$$w_p = \frac{E[\tilde{r}_p] - \frac{A}{C}}{\frac{D}{C}} V^{-1}(e - \frac{A}{C}\mathrm{I})$$

and the standard deviation of the rate of return on portfolio p is

$$\sigma(\tilde{r}_p) = \begin{cases} -\frac{E[\tilde{r}_p] - \frac{A}{C}}{\sqrt{\frac{D}{C}}}, & if \quad E[\tilde{r}_p] \le \frac{A}{C} \\ \frac{E[\tilde{r}_p] - \frac{A}{C}}{\sqrt{\frac{D}{C}}}, & if \quad \frac{A}{C} \le E[\tilde{r}_p] \end{cases}$$

The portfolio frontier of all the assets is the asymptotes of the portfolio frontier of the risky assets. Any portfolio on the portfolio frontier of all the assets involves investing everything in the riskless asset and holding an arbitrage portfolio of the risky assets - a portfolio whose weights sum to zero.

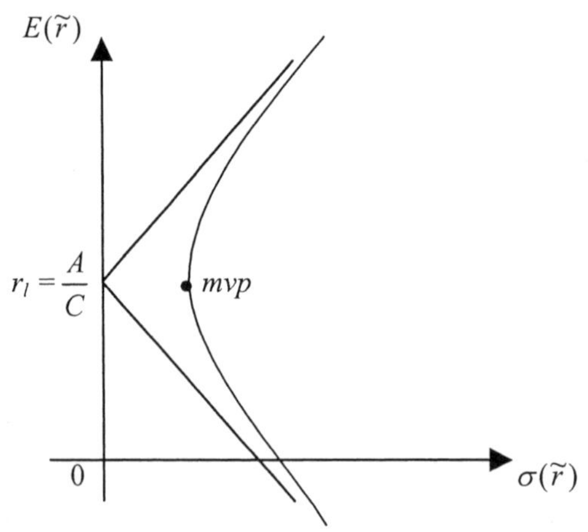

Figure 5.11 Portfolio frontier when $r_l = \frac{A}{C}$

(1.c) If $A - r_l C > 0$, then $r_l < \frac{A}{C}$. The unique set of the portfolio weights for the portfolio p having an expected rate of return of $E[\tilde{r}_p]$ is

$$w_p = \begin{cases} \frac{E[\tilde{r}_p] - r_l}{H_l} V^{-1}(e - r_l \mathrm{I}), & \text{if } E[\tilde{r}_p] \le \frac{B - r_l A}{A - r_l C} \\ \frac{1}{D} V^{-1}[(E[\tilde{r}_p]C - A)e + (B - E[\tilde{r}_p]A)\mathrm{I})], & \text{if } \frac{B - r_l A}{A - r_l C} < E[\tilde{r}_p] \end{cases}$$

and the standard deviation of the rate of return on portfolio p is

$$\sigma(\tilde{r}_p) = \begin{cases} -\frac{E[\tilde{r}_p] - r_l}{\sqrt{H_l}} & \text{if } E[\tilde{r}_p] \le r_l \\ \frac{E[\tilde{r}_p] - r_l}{\sqrt{H_l}} & \text{if } r_l \le E[\tilde{r}_p] \le \frac{B - r_l A}{A - r_l C} \\ \sqrt{\frac{C}{D}(E[\tilde{r}_p] - \frac{A}{C})^2 + \frac{1}{C}}, & \text{if } \frac{B - r_l A}{A - r_l C} < E[\tilde{r}_p] \end{cases}$$

The portfolio frontier of all the assets is composed of a half-line, a closed line segment and a curve in the $\sigma(\tilde{r}_p) - E[\tilde{r}_p]$ plane: the half-line emanating from the point $(0, r_l)$ with slope $-\sqrt{H_l}$ involving short-selling portfolio e_l and investing the proceeds in the riskless asset; the closed line segment between the point $(0, r_l)$ and the point e_l involving the proceeds in the risky assets and the riskless asset without borrowing and lending; and the curve part above the point e_l of the right-branch curve of the hyperbola with center $(0, \frac{A}{C})$ and asymptotes $E[\tilde{r}_p] = \frac{A}{C} \pm \sqrt{\frac{D}{C}} \sigma(\tilde{r}_p)$ involving short-selling (lending) the riskless asset and investing the proceeds in portfolio of the risky assets.

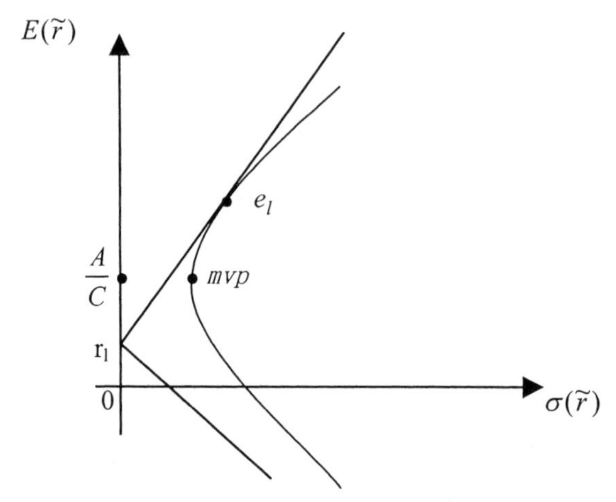

Figure 5.12 Portfolio frontier when $r_l < \frac{A}{C}$

5.3.2 Solution to (QP 2)

Now we solve quadratic program (QP2). Similarly, any w in the constraint of the quadratic program (QP2) is a regular point and we can apply the Kuhn-Tucker optimality condition to derive the unique solution of quadratic program (QP2).

Forming the Lagrangian function, we let w_p be a solution to the following unconstrained minimization problem:

$$\text{minimize } L = \frac{1}{2} w^T V w - \lambda\{w^T e + (1 - w^T \mathrm{I}) r_b - E[\tilde{r_p}]\} - \mu\{w^T \mathrm{I} - 1\}$$

where μ is a positive constant. Thus, by the Kuhn-Tucker optimality condition, we get

$$\frac{\partial L}{\partial w} = V w_p - \lambda(e - r_b \mathrm{I}) - \mu \mathrm{I} = 0 \tag{5.15}$$

$$\frac{\partial L}{\partial \lambda} = -\{w_p^T e + (1 - w_p^T \mathrm{I}) r_b - E[\tilde{r_p}]\} = 0 \tag{5.16}$$

$$w_p^T \mathrm{I} > 1 \tag{5.17}$$

$$\mu \geq 0 \tag{5.18}$$

$$\mu(w_p^T \mathrm{I} - 1) = 0 \tag{5.19}$$

From Inequality (5.17) and Equation (5.19), we have $\mu = 0$. The above conditions can be rewritten as follows:

$$Vw_p = \lambda(e - r_b \mathrm{I}) \tag{5.20}$$

$$w_p^T (e - r_b \mathrm{I}) = E[\tilde{r}_p] - r_b \tag{5.21}$$

$$w_p^T \mathrm{I} > 1 \tag{5.22}$$

Solving Equation (5.15) for w_p gives

$$w_p = \lambda V^{-1}(e - r_b \mathrm{I}) \tag{5.23}$$

which gives $w_p^T = \lambda(e - r_b \mathrm{I})^T V^{-1}$. Right-multiplying both sides by $e - r_b \mathrm{I}$ and applying (5.16), we have

$$\lambda H_b = E[\tilde{r}_p] - r_b \tag{5.24}$$

where $H_b = (e - r_b \mathrm{I})^T V^{-1}(e - r_b \mathrm{I}) = B - 2Ar_b + Cr_b^2 > 0$. Solving it for λ, we have

$$\lambda = \frac{E[\tilde{r}_p] - r_b}{H_b} \tag{5.25}$$

Combining Equations (5.23) and (5.24) gives the unique set of portfolio weights for the portfolio p to the expected rate of return of $E[\tilde{r}_b]$ and the variance of portfolio p:

$$w_p = \frac{E[\tilde{r}_p] - r_b}{H_b} V^{-1}(e - r_b \mathrm{I}) \tag{5.26}$$

and

$$\sigma^2(\tilde{r}_p) = w_p^T V w_p = \frac{(E[\tilde{r}_p] - r_b)^2}{H_b} \tag{5.27}$$

The second equality follows from Equation (5.26). Equivalently, we can write

$$\sigma(\tilde{r}_p) = \begin{cases} -\dfrac{E[\tilde{r}_p] - r_b}{\sqrt{H_b}}, & if \quad E[\tilde{r}_p] \leq r_b \\ \dfrac{E[\tilde{r}_p] - r_b}{\sqrt{H_b}}, & if \quad r_b \leq E[\tilde{r}_p] \end{cases} \tag{5.28}$$

Note that $w_p^T \mathrm{I} = \dfrac{E[\tilde{r}_p] - r_b}{H_b}(e - r_b \mathrm{I})^T V^{-1} \mathrm{I} = \dfrac{E[\tilde{r}_p] - r_b}{H_b}(A - r_b C)$. By Inequality (5.17), $(E[\tilde{r}_p] - r_b)(A - r_b C) > H_b$ and it follows that

$$E[\tilde{r}_p](A - r_b C) > B - r_b A \tag{5.29}$$

(2.a) If $A - r_b C < 0$, then $r_b > \dfrac{A}{C}$ and $E[\tilde{r}_p] < \dfrac{B - r_b A}{A - r_b C} \leq r_b$. Therefore,

$$\sigma(\tilde{r}_p) = -\frac{E[\tilde{r}_p] - r_b}{\sqrt{H_b}}$$

The portfolio frontier of all the assets is composed of a half-line emanating from the point e_b' in the $\sigma(\tilde{r}) - E[\tilde{r}]$ plane with slope $-\sqrt{H_b}$. This involves in short-selling of the riskless asset and investing the proceeds in portfolio e_b' .

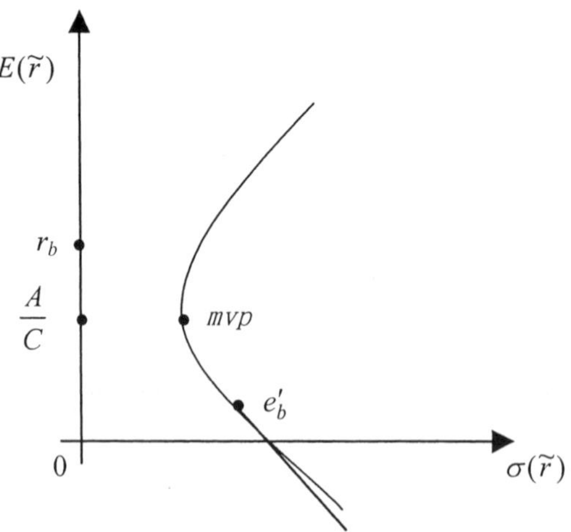

Figure 5.13 Portfolio frontier when $r_b > \dfrac{A}{C}$

(2.b) If $A - r_b C = 0$, then $r_b = \frac{A}{C}$ and Inequality (5.29) does not hold.

(2.c) If $A - r_b C > 0$, then $r_b < \frac{A}{C}$ and $E[\tilde{r_p}] > \frac{B - r_b A}{A - r_b C} \geq r_b$.

Therefore,

$$\sigma(\tilde{r_p}) = \frac{E[\tilde{r_p}] - r_b}{\sqrt{H_b}}$$

The portfolio frontier of all the assets is composed of a half-line emanating from the point e_l in the $\sigma(\tilde{r_p}) - E[\tilde{r_p}]$ plane with slope $\sqrt{H_b}$. It involves in short-selling (borrowing) the riskless asset and investing the proceeds in the portfolio e_b.

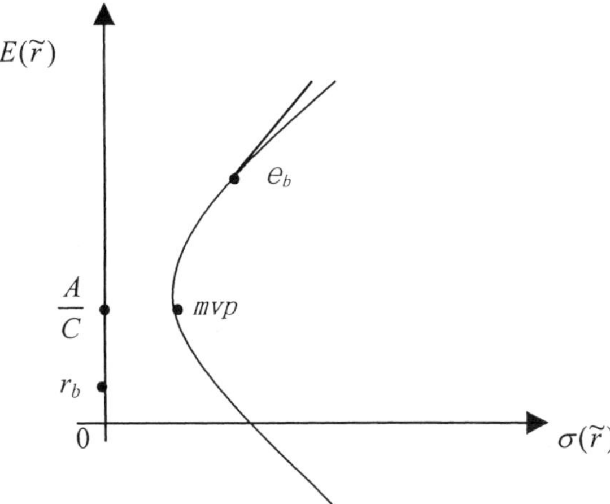

Figure 5.14 Portfolio frontier when $r_b < \frac{A}{C}$

5.3.3 Case of $r_b > r_l$

We give a few results for the quadratic program (QP) where the riskless borrowing rate is higher than the riskless lending rate, $r_b > r_l$.

(a) $\frac{A}{C} < r_l < r_b$, First we discuss point P [See Figure 5.15]. The point P is the intersection point between two half-lines, one emanating from the point $(0, r_l)$

with slope $-\sqrt{H_l}$, that is, $E[\tilde{r}_p]=r_l-\sqrt{H_l}\sigma(\tilde{r}_p)$, and another emanating from the point $(0,r_b)$ with slope $-\sqrt{H_b}$, that is, $E[\tilde{r}_p]=r_b-\sqrt{H_b}\sigma(\tilde{r}_p)$. Thus, $P(\frac{r_b-r_l}{\sqrt{H_b}-\sqrt{H_l}},\frac{\sqrt{H_b}r_l-\sqrt{H_l}r_b}{\sqrt{H_b}-\sqrt{H_l}})$.

The unique set of portfolio weights for ortfolio p having an expected rate of return of $E[\tilde{r}_p]$ is

$$w_p=\begin{cases}\frac{E[\tilde{r}_p]-r_b}{H_b}V^{-1}(e-r_b\mathrm{I}), & if \quad E[\tilde{r}_p]\leq E[\tilde{r}_P]\\ \frac{E[\tilde{r}_p]-r_l}{H_l}V^{-1}(e-r_l\mathrm{I}), & if \quad E[\tilde{r}_P]\leq E[\tilde{r}_p]\end{cases}$$

and the standard deviation of the rate of return on portfolio p is

$$\sigma(\tilde{r}_p)=\begin{cases}-\frac{E[\tilde{r}_p]-r_b}{\sqrt{H_b}}, & if \quad E[\tilde{r}_p]\leq E[\tilde{r}_P]\\ -\frac{E[\tilde{r}_p]-r_l}{\sqrt{H_l}}, & if \quad E[\tilde{r}_P]\leq E[\tilde{r}_p]\leq r_l\\ \frac{E[\tilde{r}_p]-r_l}{\sqrt{H_l}}, & if \quad r_l\leq E[\tilde{r}_p]\end{cases}$$

The portfolio frontier of all the assets is composed of two half-lines and a closed line segment in the $\sigma(\tilde{r}_p)-E[\tilde{r}_p]$ plane: the half-line emanating from point P with slope $-\sqrt{H_b}$ involving short-selling the riskless asset and investing portfolio e'_b; the closed line segment between the point $(0,r_l)$ and the point P involving investing the proceeds in the risky assets and the riskless asset without borrowing and lending; and the half-line emanating from the point $(0,r_l)$ with slope $\sqrt{H_l}$ involving short-selling portfolio e'_l and investing (lending) the proceeds in the riskless asset.

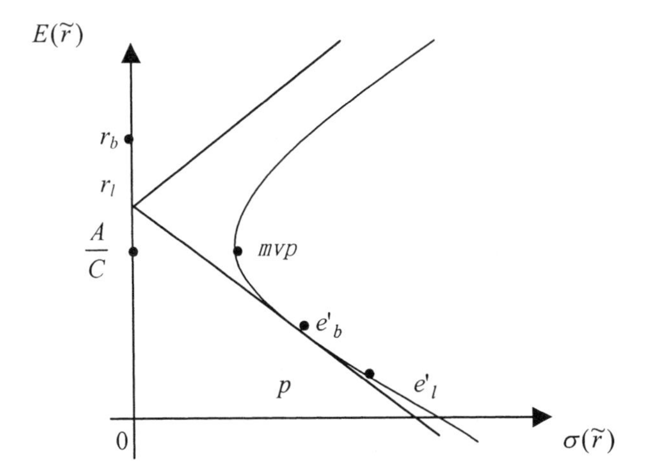

Figure 5.15 Portfolio frontier when $\frac{A}{C} < r_l < r_b$

(b) $\frac{A}{C} = r_l < r_b$. We discuss an important point denoted as P in Figure 5.16. The point P is the intersection point between two half-lines: one emanating from point $(0, \frac{A}{C})$ with slope $-\sqrt{\frac{D}{C}}$, that is, $E[\tilde{r}_p] = \frac{A}{C} - \sqrt{\frac{D}{C}}\sigma(\tilde{r}_p)$, and the other emanating from the point $(0, r_b)$ with slope $-\sqrt{H_b}$, that is, $E[\tilde{r}_p] = r_b - \sqrt{H_b}\sigma(\tilde{r}_p)$. Thus, $P\left(\frac{r_b - \frac{A}{C}}{\sqrt{H_b} - \sqrt{\frac{D}{C}}}, \frac{\sqrt{H_b}\frac{A}{C} - \sqrt{\frac{D}{C}}r_b}{\sqrt{H_b} - \sqrt{\frac{D}{C}}}\right)$.

The unique set of portfolio weights for portfolio p having an expected rate of return of $E[\tilde{r}_p]$ is

$$w_p = \begin{cases} \dfrac{E[\tilde{r}_p] - r_b}{H_b} V^{-1}(e - r_b \mathrm{I}), & if \quad E[\tilde{r}_p] \le E[\tilde{r}_P] \\ \dfrac{E[\tilde{r}_p] - \frac{A}{C}}{\frac{D}{C}} V^{-1}(e - \frac{A}{C}\mathrm{I}), & if \quad E[\tilde{r}_P] \le E[\tilde{r}_p] \end{cases}$$

and the standard deviation of the rate of return on portfolio p is

$$\sigma(\tilde{r}_p) = \begin{cases} -\dfrac{E[\tilde{r}_p]-r_b}{\sqrt{H_b}}, & if \quad E[\tilde{r}_p] \le E[\tilde{r}_P] \\ -\dfrac{E[\tilde{r}_p]-\dfrac{A}{C}}{\sqrt{\dfrac{D}{C}}}, & if \quad E[\tilde{r}_P] \le E[\tilde{r}_p] \le \dfrac{A}{C} \\ \dfrac{E[\tilde{r}_p]-\dfrac{A}{C}}{\sqrt{\dfrac{D}{C}}}, & if \quad \dfrac{A}{C} \le E[\tilde{r}_p] \end{cases}$$

The portfolio frontier of all the assets is composed of two half-lines and a closed line segment in the $\sigma(\tilde{r}_p) - E[\tilde{r}_p]$ plane: the half-line emanating from point P with slope $-\sqrt{H_b}$ involving short-selling the riskless asset and investing portfolio e'_b; the closed line segment between point $(0, \frac{A}{C})$ and point P and the half-line emanating from $(0, \frac{A}{C})$ with slope $\sqrt{\frac{D}{C}}$ involving investing everything in the riskless asset and holding an arbitrage portfolio of the risky assets - a portfolio whose weights sum to zero.

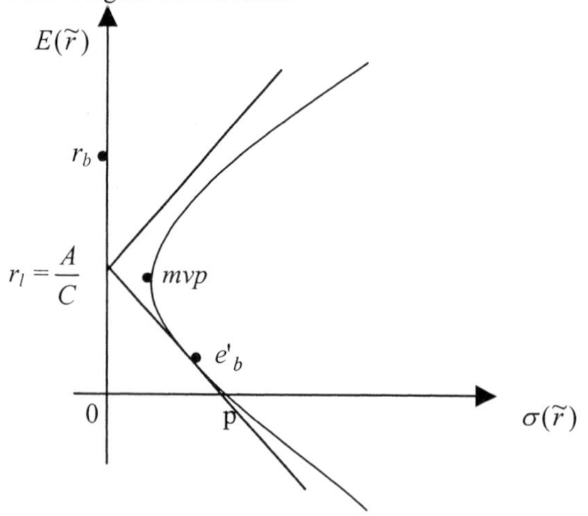

Figure 5.16 Portfolio frontier when $\frac{A}{C} = r_l < r_b$

(c) $r_l < \frac{A}{C} < r_b$. We discuss two cases: $r_b - \frac{A}{C} > \frac{A}{C} - r_l$ and $r_b - \frac{A}{C} \le \frac{A}{C} - r_l$.

(c.1) $r_b - \frac{A}{C} > \frac{A}{C} - r_l$. Thus $H_b > H_l$. Let us consider point P in Figure 5.17. The point P is the intersection point between two half-lines: one emanating from point $(0, r_l)$ with slope $-\sqrt{H_l}$, that is, $E[\tilde{r}_p] = r_l - \sqrt{H_l}\sigma(\tilde{r}_p)$, and the other emanating from point $(0, r_b)$ with slope $-\sqrt{H_b}$, that is, $E[\tilde{r}_p] = r_b - \sqrt{H_b}\sigma(\tilde{r}_p)$. Thus, $P(\frac{r_b - r_l}{\sqrt{H_b} - \sqrt{H_l}}, \frac{\sqrt{H_b} r_l - \sqrt{H} r_b}{\sqrt{H_b} - \sqrt{H_l}})$.

The unique set of portfolio weights for portfolio p having an expected rate of return of $E[\tilde{r}_p]$ is

$$w_p = \begin{cases} -\dfrac{E[\tilde{r}_p] - r_b}{H_b} V^{-1}(e - r_b \mathrm{I}), & \text{if} \quad E[\tilde{r}_p] \le E[\tilde{r}_P] \\ \dfrac{E[\tilde{r}_p] - r_l}{H_l} V^{-1}(e - r_l \mathrm{I}), & \text{if} \quad E[\tilde{r}_P] \le E[\tilde{r}_p] \le \dfrac{B - r_l A}{A - r_l C} \\ \dfrac{1}{D} V^{-1}[(E[\tilde{r}_p]C - A)e + (B - E[\tilde{r}_p]A)\mathrm{I}], & \text{if} \quad \dfrac{B - r_l A}{A - r_l C} < E[\tilde{r}_p] \end{cases}$$

and the standard deviation of the rate of return on portfolio p is

$$\sigma(\tilde{r}_p) = \begin{cases} -\dfrac{E[\tilde{r}_p] - r_b}{\sqrt{H_b}}, & \text{if} \quad E[\tilde{r}_p] \le E[\tilde{r}_P] \\ -\dfrac{E[\tilde{r}_p] - r_l}{\sqrt{H_l}}, & \text{if} \quad E[\tilde{r}_P] \le E[\tilde{r}_p] \le r_l \\ \dfrac{E[\tilde{r}_p] - r_l}{\sqrt{H_l}}, & \text{if} \quad r_l \le E[\tilde{r}_p] \le \dfrac{B - r_l A}{A - r_l C} \\ \sqrt{\dfrac{C}{D}(E[\tilde{r}_p] - \dfrac{A}{C})^2 + \dfrac{1}{C}}, & \text{if} \quad \dfrac{B - r_l A}{A - r_l C} < E[\tilde{r}_p] \end{cases}$$

The portfolio frontier of all the assets is composed of a half-line, two closed line segments and a curve in the $\sigma(\tilde{r}_p) - E[\tilde{r}_p]$ plane: the half-line emanating from the point P with slope $-\sqrt{H_b}$ involves short-selling the riskless asset and investing portfolio e'_b; the closed line segment between point $(0, r_l)$ and point P involves short-selling portfolio e_l and invests the proceeds in the riskless asset; the closed line segment between point $(0, r_l)$ and point e_l involves investing the proceeds in the risky assets and the riskless asset without borrowing and lending; and the curve part above point e_l of the right-branch curve of the hyperbola with center $(0, \frac{A}{C})$ and asymptotes $E[\tilde{r}_p] = \frac{A}{C} \pm \sqrt{\frac{D}{C}}\sigma(\tilde{r}_p)$ involves short-selling (lending) the riskless asset and invests the proceeds in the portfolio of the risky assets.

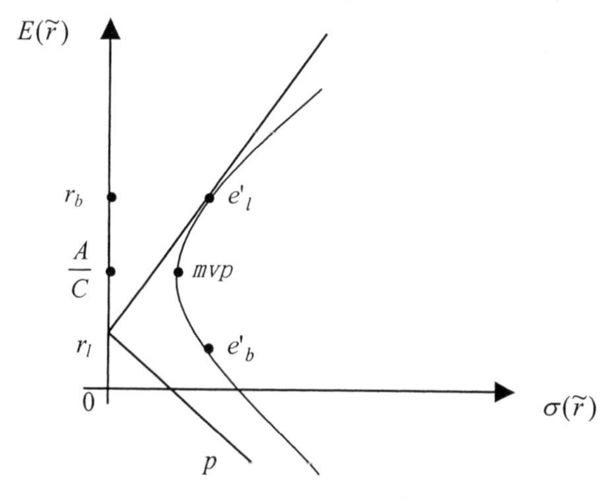

Figure 5.17 Portfolio frontier when $r_l < \frac{A}{C} < r_b$ and $r_b - \frac{A}{C} > \frac{A}{C} - r_l$

(c.2) $r_b - \frac{A}{C} > \frac{A}{C} - r_l$. The unique set of portfolio weights for optimal portfolio p having an expected rate of return of $E[\tilde{r}_p]$ is

$$w_p = \begin{cases} \dfrac{E[\tilde{r}_p] - r_l}{H_l} V^{-1}(e - r_l \mathrm{I}), & \text{if } E[\tilde{r}_p] \le \dfrac{B - r_l A}{A - r_l C} \\ \dfrac{1}{D} V^{-1}[(E[\tilde{r}_p]C - A)e + (B - E[\tilde{r}_p]A)\mathrm{I}], & \text{if } \dfrac{B - r_l A}{A - r_l C} < E[\tilde{r}_p] \end{cases}$$

and the standard deviation of the rate of return on portfolio p is

$$\sigma(\tilde{r}_p) = \begin{cases} -\dfrac{E[\tilde{r}_p] - r_l}{\sqrt{H_l}}, & \text{if} \quad E[\tilde{r}_p] \le r_l \\ \dfrac{E[\tilde{r}_p] - r_l}{\sqrt{H_l}}, & \text{if} \quad r_l \le E[\tilde{r}_p] \le \dfrac{B - r_l A}{A - r_l C} \\ \sqrt{\dfrac{C}{D}(E[\tilde{r}_p] - \dfrac{A}{C})^2 + \dfrac{1}{C}}, & \text{if} \quad \dfrac{B - r_l A}{A - r_l C} < E[\tilde{r}_p] \end{cases}$$

The portfolio frontier of all the assets is composed of a half-line, a closed line segment and a curve in the $\sigma(\tilde{r}_p) - E[\tilde{r}_p]$ plane: the half-line emanating from point $(0, r_l)$ with slope $-\sqrt{H_l}$ involves short-selling portfolio e_l and invests the proceeds into the riskless asset; the closed line segment between point $(0, r_l)$ and point e_l involves the proceeds in the risky assets and the riskless asset without borrowing and lending; and the curve part above point e_l of the right-branch curve of the hyperbola with center $(0, \frac{A}{C})$ and asymptotes $E[\tilde{r}_p] = \frac{A}{C} \pm \sqrt{\frac{D}{C}} \sigma(\tilde{r}_p)$ involves short-selling (lending) the riskless asset and invests the proceeds in the portfolio of the risky assets.

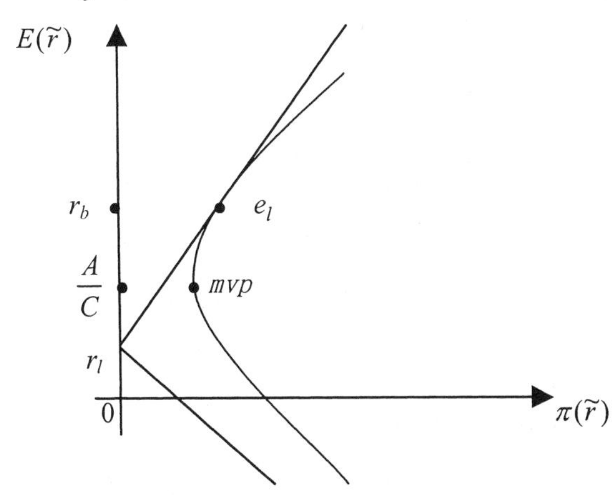

Figure 5.18 Portfolio frontier when $r_l < \frac{A}{C} < r_b$ and $r_b - \frac{A}{C} \le \frac{A}{C} - r_l$

(d) $r_l < r_b < \frac{A}{C}$. The unique set of portfolio weights for the optimal portfolio p having an expected rate of return of $E[\tilde{r}_p]$ is

$$w_p = \begin{cases} \frac{E[\tilde{r}_p]-r_l}{H_l} V^{-1}(e-r_l I), & if \quad E[\tilde{r}_p] \le \frac{B-r_l A}{A-r_l C} \\ \frac{1}{D} V^{-1}[(E[\tilde{r}_p]C-A)e+(B-E[\tilde{r}_p]A)I)], & if \quad \frac{B-r_l A}{A-r_l C} < E[\tilde{r}_p] \le \frac{B-r_b A}{A-r_b C} \\ \frac{E[\tilde{r}_p]-r_b}{H_b} V^{-1}(e-r_b I), & if \quad \frac{B-r_b A}{A-r_b C} < E[\tilde{r}_p] \end{cases}$$

and the standard deviation of the rate of return on portfolio p is

$$\sigma(\tilde{r}_p) = \begin{cases} -\frac{E[\tilde{r}_p]-r_l}{\sqrt{H_l}}, & if \quad E[\tilde{r}_p] \le r_l \\ \frac{E[\tilde{r}_p]-r_l}{\sqrt{H_l}}, & if \quad r_l \le E[\tilde{r}_p] \le \frac{B-r_l A}{A-r_l C} \\ \sqrt{\frac{C}{D}(E[\tilde{r}_p]-\frac{A}{C})^2+\frac{1}{C}}, & if \quad \frac{B-r_l A}{A-r_l C} < E[\tilde{r}_p] \le \frac{B-r_b A}{A-r_b C} \\ \frac{E[\tilde{r}_p]-r_b}{\sqrt{H_b}}, & if \quad \frac{B-r_b A}{A-r_b C} < E[\tilde{r}_p] \end{cases}$$

The portfolio frontier of all the assets is composed of two half-lines, a closed line segment and a closed curve in the $\sigma(\tilde{r}_p)-E[\tilde{r}_p]$ plane: the half-line emanating from point $(0, r_l)$ with slope $-\sqrt{H_l}$ involves short-selling portfolio e_l and invests the proceeds in the riskless asset; the closed line segment between point $(0, r_l)$ and point e_l involves short-selling portfolio e_l and invests (lends) the proceeds in the riskless asset; the closed curve part between point e_l and point e_b of the right-branch curve of the hyperbola with center $(0, \frac{A}{C})$ and asymptotes $E[\tilde{r}_p] = \frac{A}{C} \pm \sqrt{\frac{D}{C}}\sigma(\tilde{r}_p)$ involves the proceedings into the risky assets and the riskless asset without borrowing and lending; and the half-line emanating from point e_b with slope $\sqrt{H_b}$ involves short-selling (borrowing) the riskless asset and invests the proceeds in portfolio e_b.

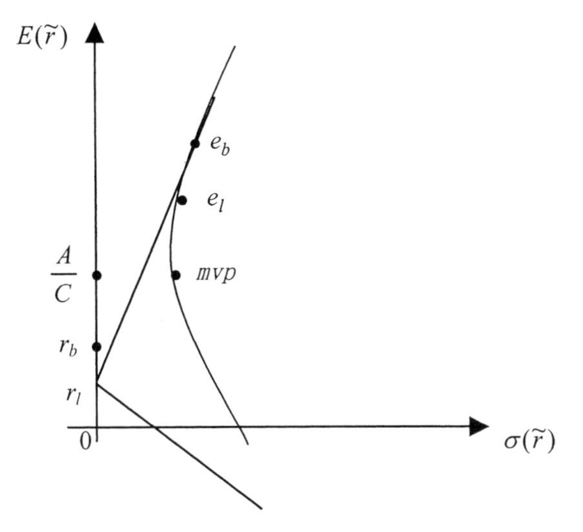

Figure 5.19 Portfolio frontier when $r_l < r_b < \dfrac{A}{C}$

5.4 Conclusions

Huang and Litzenberger (1988) studied the mathematics of the portfolio frontier. The portfolio frontier of all the risky assets is the right-branch of a hyperbola in the standard deviation--expected rate of return space while the portfolio frontier of all the risky assets and a riskless asset is two half-lines in the space.

In the new model where riskless borrowing and lending interest rates are different, the portfolio frontier of all the risky assets and a riskless asset is a continuous (smooth) curve with various simple and basic portfolio frontiers in the space. This is important for deriving various properties. For example, it follows from our analysis that there exist two (mutual) fund separation or one (mutual) fund separation. The capital asset pricing model (CAPM) also holds in this setting.

The new model, as that of Markowitz (1952), can be appled to solve the problem of two-period portfolio frontier. For a continuous time economy, Cvitanic and Karatzas (1992) developed a theory for the classical consumption/investment problem when the portfolio is constrained to take values in a given closed and convex set. They adopted an Ito process model for the financial market with one bond and a number of stocks, and studied in its framework the stochastic control problem of maximizing expected utility concerning the terminal wealth and/or consumption.

In fact, the theory of option pricing developed by Black and Scholes (1973) and Merton (1973a) is built on the idea that the option price should be equal to the cost of initiating a dynamic trading strategy in the primary assets of bond and stocks so as to guarantee the no-arbitrage condition. This approach is further generalized to consider economies with friction by Cvitani and Karatzas (1993), Karoui and Quenez (1995), and Munk (1997).

6. Multi-period Investment

6.1 Introduction

Since investors rebalance their portfolios over time, it turns out that single period investment models are not enough to help investors make their decisions about how to allocate their limited wealth to different assets. Thus multi-period models of investment are much more realistic and they are extensively used for practical purposes in the financial industry. There are some successful applications of multi-period investment models such as the Russel-Yasuda Kasai asset/liability model [Carino *et al.* (1994)], the Towers Perrin-Tillinghast asset and liability management system [Mulvey (2000)].

In this chapter, we introduce a few models for multi-period asset allocation and give some empirical comparison results about the static myopic investment models and the stochastic programming method. The first chapter of this book has introduced a number of multi-stage investment problems. The models introduced here should be considered a supplement to the first chapter. We know that it is not exhaustive.

In Section 6.2, we give some models which include the consumption and portfolio problem, the multi-period portfolio selection problems and the multi-period asset and liability management problem. Some empirical results for the multi-period stochastic programming models in finance are given in Section 6.3. Section 6.4 concludes this chapter.

6.2 Multi-period Investment Models

6.2.1. Multi-period consumption and portfolio models

The multi-period consumption and portfolio problem is studied in both discrete time environment and continuous time case.

6.2.1.1 Discrete time multi-period portfolio selection

For the discrete time case, Samuelson (1969) was among the earliest effort to study the problem. The model in the paper is to maximize the expected utility of an investor's consumption over his/her whole life period. The model is as follows:

$$J_T(W_0) = \max_{\{C_t, w_t\}} E\{\sum_{t=0}^{T}(1+\rho)^{-t}U(C_t)\}$$

$$\text{subject to } C_t = W_t - W_{t+1}[(1-w_t)(1+r)+w_t Z_t]^{-1}$$

where W_0 is the initial wealth, U is the utility function of consumption, C_t is the consumption at time t, Z_t is a random variable to represent the return of the risky asset and w_t is the proportion of money invested in this asset during time period t. r is the return of the riskless asset and ρ is the discount factor.

By the dynamic programming principle, Samuelson gave a general equation solution to the above maximization problem and found that for isoelastic marginal utility functions, the portfolio selection is independent of the consumption decision.

Hakansson (1970) presented a normative model to maximize an investor's expected utility of consumption over time. In the model, the investor has an initial wealth and also has income over time with certainty. It is assumed that he/she can lend or borrow at the riskless rate and also can put the wealth into the risky asset. The multi-period utility function is

$$U(c_1, c_2, c_3, \cdots) = u(c_1) + \alpha U(c_2, c_3, c_4, \cdots) = \sum_{j=1}^{\infty} \alpha^{j-1} u(c_j),$$

where $0 < \alpha < 1$ and is defined as the impatience level. The one period utility function can be chosen as one of the following forms:

(1) $u(c) = \frac{1}{\gamma} c^{\gamma}$, $0 < \gamma < 1$

(2) $u(c) = \frac{1}{\gamma} c^{\gamma}$, $\gamma < 0$

(3) $u(c) = \log c$,

(4) $u(c) = -e^{-\gamma c}$, $\gamma > 0$

For each of the above four utility functions, Hakansson found that the optimal consumption $c^*(x)$ is linear increasing in capital x and in non-capital income y and decreasing in α. The optimal investment strategies are independent of wealth, non-capital income and the impatience level α.

6.2.1.2 Continuous time case

Merton (1969,1971) proposed portfolio selection and consumption models in continuous time case. It is called the Merton model.

Assume that there are two assets in the market, one is bond whose return is deterministic according to the following equation: $dB_t = rB_t dt$. The other asset is stock whose price is geometric Brown motion: $dS_t = \mu S_t dt + \sigma S_t dW_t$, where W_t is standard Brown motion, μ is the drift and σ is the standard deviation. The wealth level x_t can be stated as follows:

$$dx_s = rx_s ds + (\mu - r)\pi_s ds + \sigma\pi_s dW_s$$

where π_s is the money invested in the risky asset during time period s and $x_s - \pi_s$ is the money invested in the bond asset.

Assume that the investor's initial wealth is x_0 and assume that the investor wants to maximize his/her expected utility of final wealth:

$$v(x_0, t) = \max \ E\{U(X_T) \mid x_t = x_0\}$$

When the investor's utility function is known, the optimal investment policy can be derived by the dynamic programming approach.

Interested readers are recommended to refer to [Karatzas *et al.* (1987), Karatzas *et al.* (1986), Lehoczky (1983), Pliska (1986), Davis and Norman (1990), Karatzas *et al.*(1991), Shreve and Soner (1994), Fleming and Zariphopoulou (1991), Zariphopoulou (1992, 1994, 1997), Merton (1990), Fama (1970), Pliska (1997)] for more about the consumption and investment decisions.

6.2.2 Multi-period portfolio selection models

Since the publication of Markowitz's mean-variance model, a lot of works have extended the model to multi-period models to help investors rebalance their portfolios over time.

Zhou and Li (1999) extended Markowitz's mean-variance model to the multi-period case in continuous time environment and Li and Ng (2000) presented an extension in a discrete time environment. The mathematical model for the discrete time case is as follows:

$$\text{maximize } E(x_T)$$

$$\text{subject to } \operatorname{var}(x_T) \le \sigma$$

$$x_{t+1} = \sum_{i=1}^{n} e_t^i u_t^i + (x_t - \sum_{i=1}^{n} u_t^i) e_t^0 ,\ t = 0,1,2,\cdots,T-1$$

or

$$\text{maximize } E(x_T) - w \operatorname{var}(x_T)$$

$$\text{subject to } x_{t+1} = \sum_{i=1}^{n} e_t^i u_t^i + (x_t - \sum_{i=1}^{n} u_t^i) e_t^0 ,\ t = 0,1,2,\cdots,T-1$$

where $0 \le w < \infty$ is used to represent the risk attitude of the investor, the larger the w, the more risk averse the investor is. x_T is the final wealth at the end of period T. x_t is wealth at the end of period t and u_t^i is the money invested in the risky asset i. For the above models, the authors gave an analytical result in calculating an optimal portfolio and interested readers can refer to the paper.

For more about multi-period portfolio selection models, interested readers can refer to Brennan *et al.* (1997), Markowitz (1976), Elton and Gruber (1974), Smith (1967), Mossin (1968), Chen *et al.* (1971), Winkler and Barry (1975), Dumas and Luciano (1991), and Ostermark (1991).

6.2.3 Multi-period asset and liability management

Multi-period asset and liability models are the most successful used models in practice and a lot of publications are contributed to this topic. For example, one can refer to Kusy and Ziemba (1986), Hiller and Eckstein (1991), Carino *et al.* (1994), Mulvey *et al.* (2000), Zenios *et al.* (1998), and Mulvey and Vladimirou (1992).

The most successful model of multi-period asset and liability management might be the model of Russell-Yasuda Kasai financial planning model. In the late eighties of the last century, the company's complex liability structure and casualty insurance business and the large number of restrictions in asset management motivate a new model to replace the single period Markowitz mean-variance model. Carino and Ziemba (1998) designed a multi-stage stochastic linear programming model to maximize the company's long run wealth level as well as to cover the liability. After the company implemented the new model, it yielded extra income of 42 basis points (US $79 million) over the first two years (1991 and 1992). Other successful applications include Mulvey *et al.* (2000), and Seshadri *et al.* (1999).

In the next section, we will give some empirical results as illustruation to show that the multi-stage stochastic programming methods are indeed better than the myopic single period balance model. We hope that this will motivate more academics and practitioners to explore the financial applications of stochastic programming models .

6.3 Empirical Results for Multi-period Stochastic Programming Models

Different strategies for allocating the funds to different assets are applied by different investors with different preferences. For example, Perold and Sharpe (1988) studied four classes of dynamic strategies: buy-and-hold, constant mix, constant-proportion portfolio insurance, and option based portfolio insurance. They found that it is not easy to see which strategy is best for everyone. In fact, only after the financial analysts have good knowledge of the investors' preferences, they might help investors to choose their best dynamic strategy.

In this section, we examine the static myopic model which is based on one–period optimal solution and compare it with the stochastic programming methods. Although some multi-period problems with special utility functions can be reduced to a one period problem [Ziemba and Vickson (1975)], one can give detailed suggestions about how to allocate their money to different assets only when knowing the investors' multi-period preference functions.

6.3.1 Discrete multi-period asset allocation problem

Consider two types of investments: stocks and bonds. The returns on these investments are assumed to be random variables with known distributions. For an initial wealth w_0, the purpose of the investment is to exceed a specified goal G at the end of the period T.

For a piece-wise linear utility function, denote T as the number of investment period; ρ as the unit shortage penalty, $r_{st}^{w_T}$ the return on stocks during period t

with scenario w_t, $r_{bt}^{w_t}$ the return on bonds during period t with scenario w_t. We define decision variables as follows, $s_t^{w_{t-1}}$ is the investment in stocks in time period t in scenario w_{t-1}, $b_t^{w_{t-1}}$ is the investment in bonds in time period t in scenario w_{t-1}, u^{w_T} is the shortage of target G in scenario w_T, and v^{w_T} is the excess of the target in scenario w_T. Then the discrete multi-period stochastic portfolio allocation problem with a piece-wise linear utility function is formulated as follows:

$$\text{maximize } z = \sum_{w_T \in \Omega_T} p_T^{w_T} (v^{w_T} - \rho u^{w_T})$$

$$\text{subject to } s_1 + b_1 = w_0$$

$$s_t^{w_{t-1}} + b_t^{w_{t-1}} - r_{st-1}^{w_{t-1}} s_{t-1}^{a(w_{t-1})} - r_{bt-1}^{w_{t-1}} b_{t-1}^{a(w_{t-1})} = 0 \quad \forall\, w_t \in \Omega_t \text{, t=2, ... , T}$$

$$v^{w_T} - u^{w_T} - r_{sT}^{w_T} s_T^{a(w_T)} - r_{bT}^{w_T} b_T^{a(w_T)} = -G \quad \forall\, w_T \in \Omega_T$$

$$s_t^{w_{t-1}},\ b_t^{w_{t-1}},\ u^{w_T},\ v^{w_T} \ \geq\ 0 \ \ \forall\, w_t \in \Omega_t \text{, t=2, ... , T}$$

The following is the figure of the utility function.

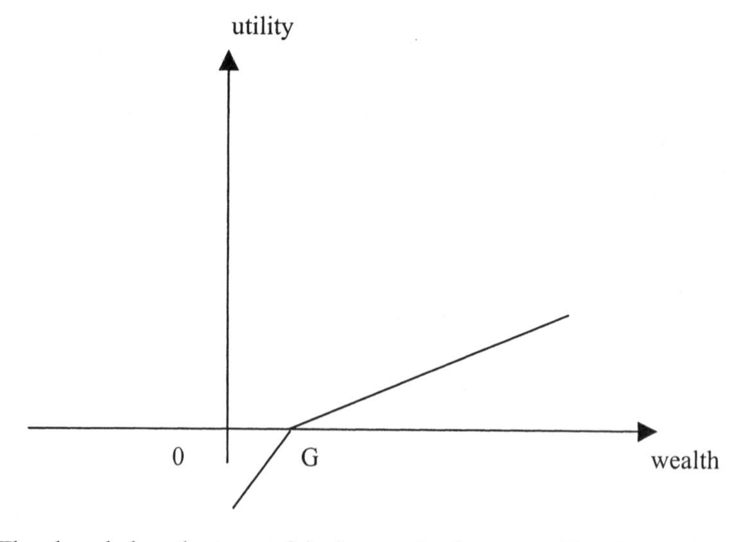

The slope below the target G is the penalty factor ρ. If ρ is equal to one, then it means risk neutrality. If ρ is less than one, it means that the investor is risk seeking. He/she concerns more about return than risk. And if ρ is greater than one, it means risk averse. The larger the value of ρ is, the more the risk averse of the investor is.

Assume that the stocks have two returns; one is 1.25, and the other is 1.06 and each is with probability 0.5. Assumed that the bonds have two types of returns, the higher return is 1.14 while the lower return is 1.12, each with probability 0.5 [Morton (2000)].

For the static model, because the expected return of stocks exceed that of bonds: (1.25+1.06)/2 > (1.14+1.12)/2, the solution is obvious, to invest the money in stocks always.

The expected wealth at the end of each period is listed in the following table:

Table 6.1 The expected wealth at the six periods.

Periods	2 period	3 period	4 period	5 period	6 period
Expected wealth at the end of the period	73.3714	84.7439	97.8792	113.0505	130.5734

Note: initial wealth = 55.

For the stochastic programming method, we fix the penalty factor as 4 and change the number of time periods. The utility benefit of different period models are listed in the following table.

Table 6.2 Utility benefit of different period moldels.

# of periods	Target wealth[2]	Static model		Stochastic method		Wealth change[3]	Utility benefit[4]
		Wealth[1]	Utility	Wealth[1]	Utility		
2 periods	70.00	73.37	-2.78	70.453	-0.468	- 4.0%	83.2%
3 periods	80.00	84.74	-3.79	83.046	-1.514	- 2.0%	60.0%
4 periods	91.00	97.88	-4.00	95.279	-1.783	- 2.7%	55.4%
5 periods	105.00	113.05	-5.91	111.392	-4.172	- 1.5%	29.4%
6 periods	120.00	130.57	-7.36	128.167	-4.950	- 1.8%	32.7%

Note:

1. The wealth means the final wealth at the end of the corresponding period, e.g., 4 periods is the final wealth at the end of the fourth period.

2. We have the target wealth a little lower than the expected wealth.
3. Wealth change means the proportion change of the final wealth of the stochastic programming method relative to that by the static model method.
4. Utility benefit means the proportion change of the utility value by the stochastic programming method relative to that by the static model method.

A chart for the utility benefit with number of periods is as follows,

Figure 6.1 Utility benefit with number of periods.

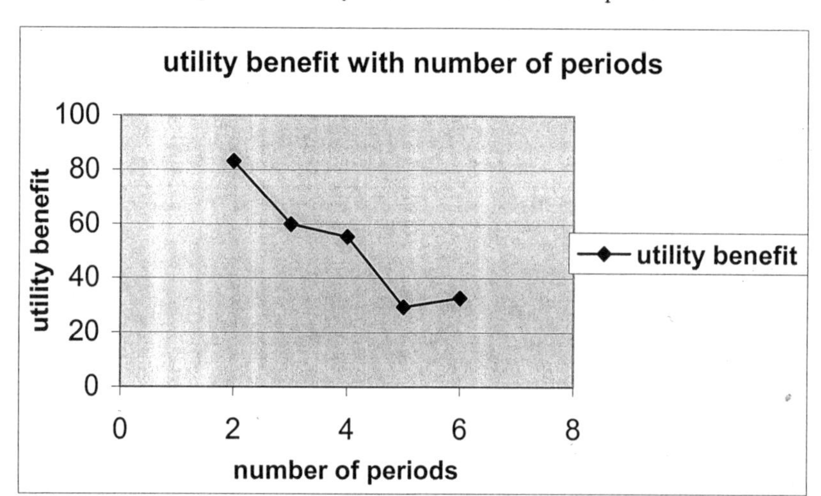

The benefit of the stochastic models are obvious, with a proportion range from 29.4% to 83.2%.

Next, we compare four models which are "2 stage update", "3 stage update", "6 stage update" and static model respectively. The first "2 stage update" model means that we update the investment allocation decision every two stages; After the investment decision at the beginning, the investor will adjust the portfolio at the end of the second stage and at the end of the fourth stage. For the "3 stage update", the investor adjusts the allocation at the end of the third stage. The "6 stage update" model means to allocate the asset as a whole six period model.

We fix the target wealth at the end of the sixth stage as 120 and change the penalty factors to derive the following results:

Table 6.3 Final wealth of different models with different penalty factors (target wealth is 120)

Penalty factor	2 stage update	3 stage update	6 stage update	Static model
6	121.29	121.76	124.63	130.57
4	122.12	124.84	128.17	130.57
2	127.69	129.62	130.01	130.57
1.5	130.57	130.57	130.57	130.57
1.0	130.57	130.57	130.57	130.57
0.8	130.57	127.53	127.39	130.57
0.5	130.57	127.53	124.55	130.57
0.2	126.04	126.15	125.76	130.57
0.0	122.28	126.15	125.76	130.57

Figure 6.2 Final wealth of different stage update and different penalty factor.

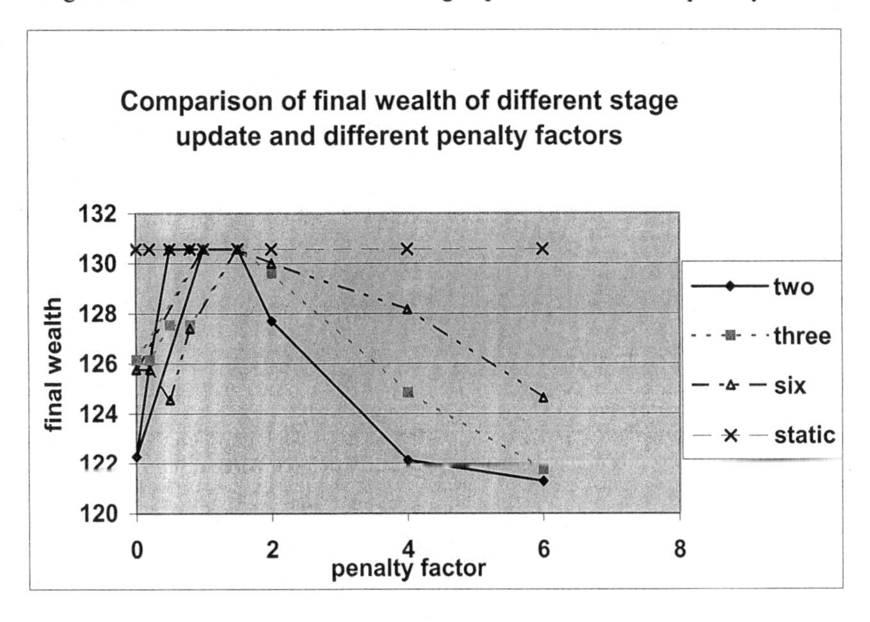

Table 6.4 Utility value of different models with different penalty factors

(target wealth is 120)

Penalty factor	2 stage update	3 stage update	6 stage update	Static model
6	-0.04	-5.56	-12.79	-19.31
4	0.55	-1.68	-4.95	-7.36
2	5.81	6.98	83.35	4.60
1.5	9.32	9.20	7.59	7.58
1.0	10.57	10.57	10.57	10.57
0.8	11.08	8.16	8.60	11.77
0.5	11.83	9.10	7.63	13.56
0.2	8.56	8.53	10.57	15.36
0.0	6.62	9.13	11.77	16.55

Figure 6.3 Utility of different stage updates and different penalty factor.

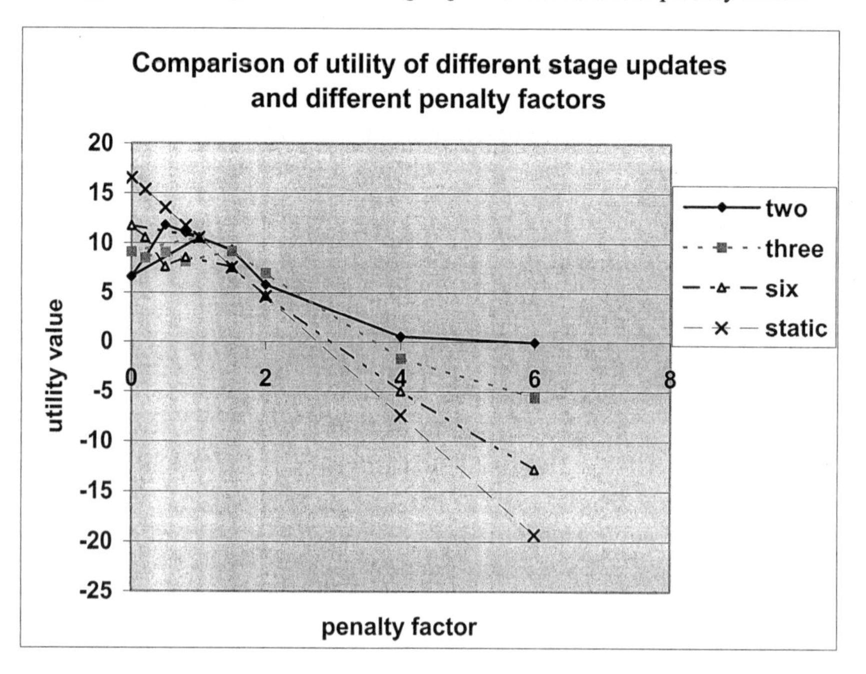

From the above tables and figures, we can conclude that, for the multi-period stochastic models, the expected utility value for the risk averseness behavior is larger than the static model, the solution is the same for risk neutrality behavior and for the risk seeking behavior, and the expected utility of the stochastic models is less than those results from the static model.

More specifically, we shall consider different risk attitude in the following subsections.

6.3.1.1 Risk averseness behavior results

For the risk averseness behavior, we consider different target values at the end of the six stage and have the following utility value table:

Table 6.5 Utility value of risk averse behavior with different target values (penalty factor = 2)

Target value	2 stage update	3 stage update	6 stage update	Static model
150	-39.52	-41.42	-42.60	-42.61
140	-23.36	-23.56	-25.89	-26.05
120	5.81	6.98	4.71	4.60
100	29.18	30.49	29.54	29.48
80	50.01	50.25	50.55	50.54

Figure 6.4 Utility of different stage updates and the static model

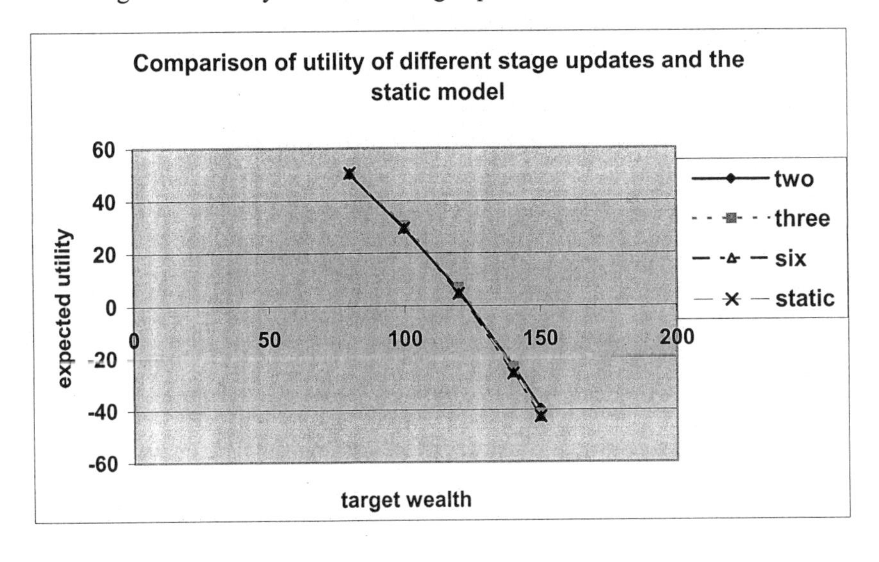

We can observe that the stochastic model is definitely better than the static model, however, it is difficult is see when to adjust the asset decisions. For this problem data set, the "3 stage update" model seems better.

Table 6.6 Final wealth of risk averse behavior with different target values (penalty factor = 2)

Target value	2 stage update	3 stage update	6 stage update	Static model
150	130.24	129.32	130.51	130.57
140	128.32	129.69	129.80	130.57
120	127.69	129.62	130.01	130.57
100	129.18	130.49	130.27	130.57
80	130.01	130.25	130.55	130.57

Figure 6.5 Final wealth with different stage updates and the static model

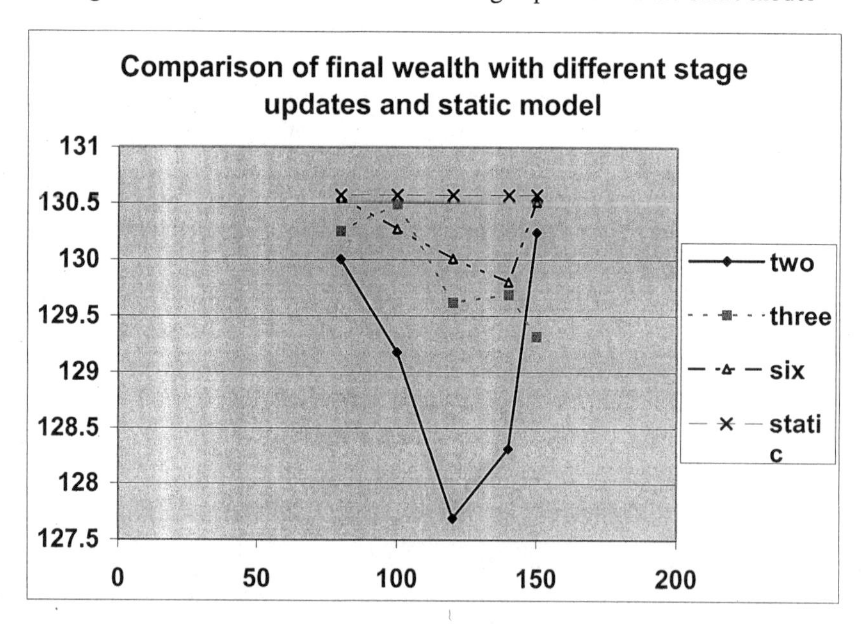

One can observe that the final wealth of the different stochastic models are less than the static model.

6.3.1.2 Risk seeking behavior results

For the risk seeking behavior, when the penalty factor is equal to 0.5, we have the following table for the final wealth after six periods:

Table 6.7 Final wealth of risk seeking behavior with different target values (penalty factor 0.5)

Target value	2 stage update	3 stage update	6 stage update	Static model
150	130.57	127.78	125.76	130.57
140	130.57	126.15	125.76	130.57
120	130.57	127.53	125.55	130.57
100	130.57	126.25	124.15	130.57
80	122.28	121.05	124.15	130.57

The corresponding figure is as follows:

Figure 6.6 Final wealth for the risk seeking behavior.

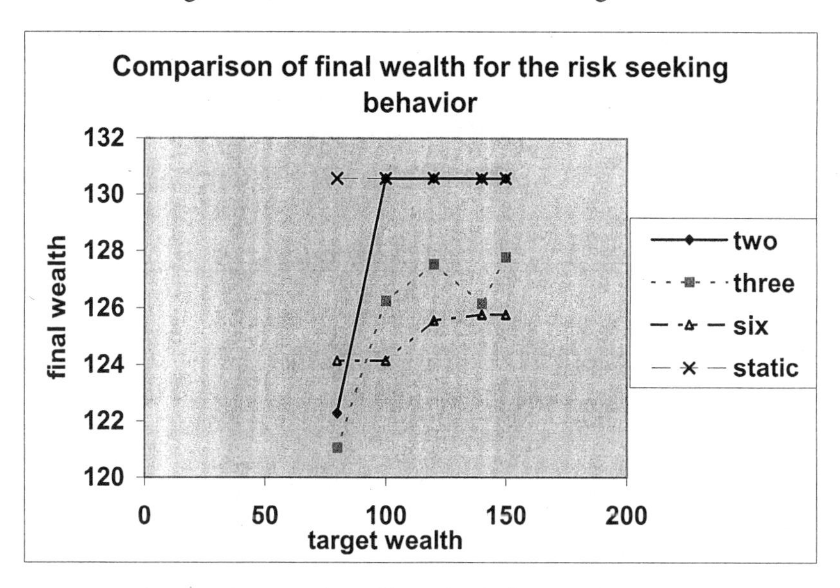

At the end of the sixth stage, the utility value is listed below:

Table 6.8 Utility value of risk seeking behavior with different target values (penalty factor = 0.5)

Target value	2 stage update	3 stage update	6 stage update	Static model
150	-9.35	-10.25	-11.27	-7.84
140	-3.10	-6.04	-5.44	-1.12
120	11.83	9.10	7.63	13.56
100	30.57	26.25	24.48	31.12
80	42.28	41.05	44.15	50.59

Figure 6.7 Utility for different stage updates and different target values

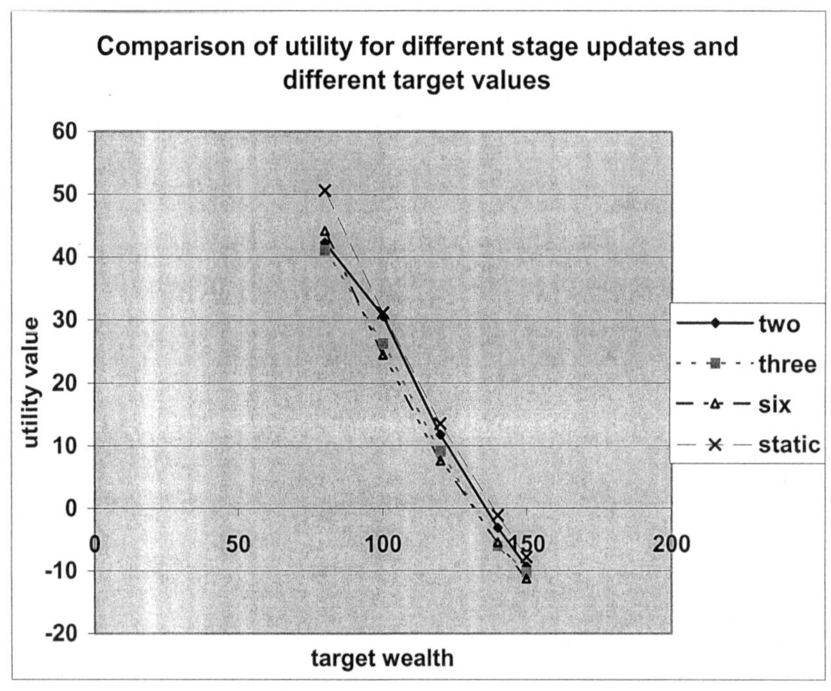

Among all the four models, the static model has the highest utility value.

6.3.2 Continuous multi-period asset allocation problem

For the continuous multi-period asset allocation problem, we use the same data as in Ziemba *et al.* (1974). We assume that the assets are normally distributed. The yearly data and quarterly data are listed in the following two tables:

Table 6.9 Yearly data

Name	Return	Variance-covariance matrix			
CT	1.098367	0.013683	0.014316	0.014347	0.004608
MT	1.094887	0.014316	0.027606	0.022203	0.006393
RT	1.094135	0.014347	0.022203	0.021802	0.006586
NT	1.087323	0.004608	0.006393	0.006586	0.011349

Table 6.10 Quarterly data.

Name	Return	Variance-covariance matrix		
CT	1.022642	0.004471	0.003431	0.003403
MT	1.021878	0.003431	0.003769	0.003577
RT	1.020697	0.003403	0.003577	0.004230

Note: CT is Canada Trust, MT is Montreal Trust, NT is National Trust and RT is Royal Trust. Since NT is published only yearly so there are only three assets for the quarterly data.

As shown in Ziembia *et al.* (1974), the static optimal portfolio is derived by solving the following programming problem:

$$\text{maximize} \quad \frac{rx}{(x'\Sigma x)^{\frac{1}{2}}}$$

$$\text{subject to} \quad \sum x_i = 1$$

$$x_i \geq 0, \quad \forall i$$

For the yearly data, the optimal solution of the above problem is: $x_Y^* =$ (1.000,0.000, 0.000, 0.000). The composite risky asset is normally *N(1.098,*

0.0137). For the quarterly data, its optimal solution is $x_Q^* = (0.500,0.500,0.000)$, the composite asset is normally distributed with mean 1.022 and variance 0.00378. For the stochastic optimization methods, we consider both its upper bound and lower bound.

Consider the general utility maximization problem:

$$\max_{x \in X} \quad EU(rx) \tag{P1}$$

Denote $r^1, r^2, \cdots, r^m$ as *i.i.d* random variables from $r = (r_1, r_2, r_3, r_4)$, the above problem is equivalent as the following problem:

$$\max_{x \in X} \quad E(\frac{1}{m}\sum_{i=1}^{m} U(r^i x))$$

From the Jensen inequality, an upper bound is given by the following problem:

$$\text{E}[\max_{x \in X} \quad \frac{1}{m}\sum_{i=1}^{m} U(r^i x)]$$

Denote x' as a feasible solution of *(P1)*. Thus a lower bound is given by

$$L(n_L) = \frac{1}{n_L}\sum_{i=1}^{n_L} U(r^i x')$$

We adopt $n_U = n_L = 20$ as the number of sample batches and m=30 as the number of points within a batch. We use common random generator methods, in the following, we give two 95% confidence intervals for the optimality gap between the upper bound and the lower bound. The lower bound is given differently for the two confidence intervals. One lower bound is given by a feasible solution derived from the upper bound while the other is the optimal solution of the static model in Ziemba *et al.* (1974).

Since the size of the continuous stochastic multi-period asset allocation problem by sampling is huge, we only consider the two period model and the three period model. We give some results with several different utility functions.

6.3.2.1 Piece-wise linear utility function

The piece-wise linear utility function is given by

$$u(v, y) = y - \rho v$$

where v is the shortage of the target value at the end of the investment period, y is the excess value of the target and ρ is the penalty factor.

Table 6.11 Comparison of confidence intervals (penalty factor = 4.0)
(Yearly data)

Model	Target value	Stochastic Solution	Confidence interval one	Confidence interval two	Range deduction
2 period	1.206	(0.310, 0.204, 0.326, 0.160)	(0, 0.064)	(0, 0.040)	37.5%
3 period	1.324	(0.255, 0.221, 0.358, 0.165)	(0, 0.139)	(0, 0.066)	52.5%

Note: Confidence interval one is derived by the static model optimal solution while confidence interval two is from the feasible solution by solving the stochastic optimization model's upper bound.

Table 6.12 Comparison of confidence intervals (penalty factor = 4.0)
(Quarterly data)

Model	Target value	Stochastic Solution	Confidence interval one	Confidence interval two	Range deduction
2 period	1.044	(0.381, 0.385, 0.234)	(0, 0.023)	(0, 0.015)	34.8%
3 period	1.067	(0.264, 0.330, 0.407)	(0, 0.022)	(0, 0.015)	31.8%

Consider the confidence intervals with different penalty factors for the two-period stochastic asset allocation problem, we have the following results:

Table 6.13 Comparison of confidence intervals with different penalty factors (target = 1.044) (Yearly data)

Penalty factor	Stochastic solution	Confidence interval one	Confidence Interval two
6	(0.317, 0.201, 0.302, 0.181)	(0, 0.103)	(0, 0.051)
4	(0.310, 0.204, 0.326, 0.160)	(0, 0.064)	(0, 0.040)
2	(0.306, 0.256, 0.351, 0.087)	(0, 0.032)	(0, 0.035)

1.5	(0.307,0.306, 0.344, 0.043)	(0, 0.028)	(0, 0.035)
1.0	(0.200,0.450, 0.350, 0.000)	(0, 0.024)	(0, 0.039)
0.8	(0.250,0.400, 0.350, 0.000)	(0, 0.024)	(0, 0.036)
0.5	(0.300,0.350, 0.350, 0.000)	(0, 0.023)	(0, 0.034)
0.2	(0.323,0.354, 0.323, 0.000)	(0, 0.025)	(0, 0.046)
0.0	(0.250,0.350, 0.400, 0.000)	(0, 0.023)	(0, 0.038)

Table 6.14 Comparison of confidence intervals with different penalty factors (target = 1.206) (Quarterly data)

Penalty factor	Stochastic solution	Confidence interval one	Confidence interval two
6	(0.364, 0.385, 0.249)	(0, 0.033)	(0, 0.019)
4	(0.381, 0.385, 0.234)	(0, 0.023)	(0, 0.015)
2	(0.451, 0.351, 0.198)	(0, 0.014)	(0, 0.013)
1.5	(0.504, 0.335, 0.161)	(0, 0.013)	(0, 0.011)
1.0	(0.550, 0.350, 0.100)	(0, 0.013)	(0, 0.012)
0.8	(0.500, 0.350, 0.150)	(0, 0.011)	(0, 0.012)
0.5	(0.500, 0.300, 0.200)	(0, 0.011)	(0, 0.013)
0.2	(0.521, 0.279, 0.200)	(0, 0.010)	(0, 0.012)
0.0	(0.450, 0.350, 0.200)	(0, 0.009)	(0, 0.012)

6.3.2.2 Quadratic utility function

The quadratic utility function is given by

$$u(w) = w - \beta w^2$$

This function is concave when β is positive. The larger the β value, the more risk averseness the investor is. Consider the confidence intervals with different penalty factors for the two-period stochastic asset allocation problem, we have the following results:

Table 6.15 Comparison of confidence intervals (Yearly data)

β	Solution	Confidence interval one	Confidence interval two
0.5	(0.100, 0.200, 0.150, 0.550)	(0, 0.008)	(0, 0.004)
1	(0.100, 0.200, 0.150, 0.550)	(0, 0.071)	(0, 0.040)
1.5	(0.100, 0.200, 0.150, 0.550)	(0, 0.132)	(0, 0.077)
2	(0.100, 0.200, 0.150, 0.550)	(0, 0.194)	(0, 0.111)
4	(0.100, 0.200, 0.150, 0.550)	(0, 0.441)	(0, 0.256)
6	(0.100, 0.200, 0.150, 0.550)	(0, 0.401)	(0, 0.689)

Table 6.16 Comparison of confidence intervals (Quarterly data)

β	Solution	Confidence interval one	Confidence interval two
0.5	(0.100, 0.400, 0.500)	(0, 0.001)	(0, 0.001)
1	(0.100, 0.400, 0.500)	(0, 0.018)	(0, 0.014)
1.5	(0.100, 0.400, 0.500)	(0, 0.38)	(0, 0.029)

2	(0.100, 0.400, 0.500)	(0, 0.56)	(0, 0.042)
4	(0.100, 0.400, 0.500)	(0, 0.129)	(0, 0.098)
6	(0.100, 0.400, 0.500)	(0, 0.203)	(0, 0.154)

6.3.2.3 Cubic utility function

The cubic utility function is given by

$$u(w) = w^3 - 2kw^2 + (k^2 + g^2)w$$

If $0 \le w \le \frac{2}{3}k - \frac{1}{3}(k^2 - 3g^2)^{1/2}$, the utility function is concave. Consider the confidence intervals with different penalty factors for the two-period stochastic asset allocation problem, we have the following results:

Table 6.17 Comparison of confidence intervals (Yearly data)

k and g value	Solution	Confidence interval one	Confidence interval two
G=0.5 k=4	(0.200, 0.450, 0.350, 0.000)	(0, 0.026)	(0, 0.040)
G=1 k=4.5	(0.200, 0.450, 0.350, 0.000)	(0, 0.087)	(0, 0.133)
G=1.5 k=5	(0.200, 0.450, 0.350, 0.000)	(0, 0.175)	(0, 0.265)
G=2 k =5	(0.200, 0.450, 0.350, 0.000)	(0, 0.218)	(0, 0.331)
G= 2.5 k=5.5	(0.200, 0.450, 0.350, 0.000)	(0, 0.340)	(0, 0.519)
G=1 k = 3.5	(0.200, 0.450, 0.350, 0.000)	(0, 0.012)	(0, 0.021)

Table 6.18 Comparison of confidence intervals (Quarterly data)

k and g value	Solution	Confidence interval one	Confidence interval two
G=0.5 k=4	(0.550, 0.350, 0.100)	(0.000,0.032)	(0.000,0.031)
G=1 k=4.5	(0.550, 0.350, 0.100)	(0.000,0.068)	(0.000,0.067)
G=1.5 k=5	(0.550, 0.350, 0.100)	(0.000,0.113)	(0.000,0.114)
G=2 k =5	(0.550, 0.350, 0.100)	(0.000,0.133)	(0.000,0.133)
G= 2.5 k=5.5	(0.550, 0.350, 0.100)	(0.000,0.199)	(0.000,0.197)
G=1 k = 3.5	(0.550, 0.350, 0.100)	(0.000,0.068)	(0.000,0.067)

6.3.2.4 Exponential utility function

The exponential utility function is given by

$$u(w) = 1 - e^{-aw}$$

It is concave if $a \geq 0$. Consider the confidence intervals with different penalty factors for the two-period stochastic asset allocation problem, we have the following results:

Table 6.19 Comparison of confidence intervals (Yearly data)

A	Solution	Confidence interval one	Confidence interval two
0.5	(0.200, 0.450, 0.350, 0.000)	(0,0.006)	(0,0.010)
1	(0.200, 0.450, 0.350, 0.000)	(0,0.007)	(0,0.010)
1.5	(0.200, 0.450, 0.350, 0.000)	(0,0.006)	(0,0.009)

2	(0.200, 0.450, 0.350, 0.000)	(0,0.005)	(0,0.007)
4	(0.200, 0.450, 0.350, 0.000)	(0, 0.001)	(0, 0.002)
6	(0.200, 0.450, 0.350, 0.000)	(0, 0.000)	(0, 0.000)
8	(0.200, 0.450, 0.350, 0.000)	(0, 0.000)	(0, 0.000)
15	(0.200, 0.450, 0.350, 0.000)	(0, 0.000)	(0, 0.000)

Table 6.20 Comparison of confidence intervals (Quarterly data)

a	Solution	Confidence interval one	Confidence interval two
0.5	(0.550, 0.350, 0.100)	(0.000,0.004)	(0.000,0.003)
1	(0.550, 0.350, 0.100)	(0.000,0.004)	(0.000,0.003)
1.5	(0.550, 0.350, 0.100)	(0.000,0.004)	(0.000,0.003)
2	(0.550, 0.350, 0.100)	(0.000,0.003)	(0.000,0.003)
4	(0.550, 0.350, 0.100)	(0.000,0.001)	(0.000,0.001)
6	(0.500, 0.350, 0.150)	(0.000,0.000)	(0.000,0.000)
8	(0.500, 0.350, 0.150)	(0.000,0.000)	(0.000,0.000)
15	(1.000, 0.000, 0.000)	(0.000,0.000)	(0.000,0.000)

From the above tables, we observe that for most of the times, the stochastic programming methods are better in the sense that it reduces the gap between the lower bound and the upper bound.

6.4 Conclusions

In this chapter, we discussed the stochastic multi-period asset allocation problem and have given some empirical results. We have presented results concerning random return data for both the discrete case and the continuous case. For the discrete case, the solution of the stochastic model is definitely better than the solution from the static myopic model. For the continuous case, we assume that the return of the asset is normally distributed which is not a very strong assumption [Kallberg and Ziembia (1983)].

For the multi-period asset allocation problem, the dynamic programming approach is very important from the theoretic point of view. However, it is rarely used in practice due to its complexity and time consumed for the calculation. The most useful tool for the multi-period investment problem is linear stochastic programming models. But as pointed by Mulvey (2001), it is very difficult in the sense that the variables and scenarios are enormous when there are many periods. But we believe as the efforts of both the researchers and the investors, efficient algorithms for solving the multi-period investment models will not be far away. And more and more investors will find the benefits of the multi-period models in the future.

7. Mean-Variance-Skewness Model for Portfolio Selection with Transaction Costs

7.1 Introduction

The mean-variance methodology originally proposed by Markowitz (1952) plays a crucial role in the theory of portfolio selection and gains widespread acceptance as a practical tool for portfolio optimization. Since the seminal works of Markowitz, numerous studies on portfolio selection and performance measures have been made based on only the first two moments of return distributions. However, there is a controversy over the issue of whether higher moments should be accounted for in portfolio selection. Many authors (e.g., [Arditti (1967,1971), Samuelson (1958), Rubinstein (1973), Konno *et al.* (1995)]) argued that the higher moments can not be neglected unless there is a reason to believe that the asset returns are normally distributed and the utility function is quadratic, or that the higher moments are irrelevant to the investor's decision.

In fact, there is ample empirical evidence (e.g., [Fama (1965), Arditti (1971), Simkowitz and Beedles (1978), Konno and Suzuki (1992), Konno *et al.* (1993), Pornchai *et al.* (1997)]) indicating that individual security and portfolio returns are not normally distributed. Hanoch and Levy (1970) pointed out that the quadratic utility function implies increasing absolute risk aversion which is contrary to the normal assumption of decreasing absolute risk aversion. Levy and Sarnat (1970) mentioned that the assumption of a quadratic utility function is appropriate only for relatively low returns, which precludes its use for many types of investments. Samuelson (1958) showed that the higher moment is relevant to the investor's decision in portfolio selection. Furthermore, almost every investor would prefer a portfolio with larger third moment if the first and second moments are the same. All the discussions above explain some reasons of adding the third moment of portfolio returns into a general mean-variance model, even though it may make the optimization problem more difficult and more complicated to be solved.

The mean-variance-skewness model has been studied by a number of authors, see, e.g., [Konno *et al.* (1995)]. But no investigation has been made on any mean-variance-skewness model for portfolio selection with transaction costs. However, the transaction costs are a source of constant concern for portfolio managers. Obviously, the transaction costs have a direct impact on one's investment performance. The net return of a portfolio of securities should be evaluated by taking the transaction costs into a consideration.

A number of publications (e.g., [Mao (1970b, 1970c), Pogue (1970), Chen, Jen and Zionts (1971), Jacob (1974), Brennan (1975), Levy (1978), Patel and Subhmanyam (1982), Mulvey and Vladimiron (1992), Dantzig and Infanger (1993), Gennotte and Jung (1994), Yoshimoto (1996), Li, Wang and Deng (2000)]) discussed the transaction costs in portfolio optimization. Most of them incorporated the transaction costs into the single-period or multi-period portfolio selection models. Very recently, Li, Wang and Deng (2000) gave a linear programming algorithm to solve a general mean-variance model for portfolio selection with transaction costs. Its main ideas are new and constructive. The discussion in this chapter is similar to the model of Li, Wang and Deng (2000) assuming that the applications of the portfolio optimization problem usually involve the revision of an existing portfolio, probably, due to changes in information about these securities. The transaction cost is assumed to be a V-shaped function of the difference between a new portfolio and the existing one, since the revision entails both purchases and sales of securities that for both transaction costs are paid.

The mean-variance-skewness model with transaction costs in this chapter is formed as a three-objective nonlinear programming problem. In order to solve a Pareto efficient solution, we transform it into a parametric programming problem by maximizing the skewness under given levels of the mean and the variance. We give an approximation of replacing the term "variance" by "absolute deviation" and the term "skewness" by the expectation of a piecewise linear function so that the problem can be solved by a linear programming algorithm. This implies that the method proposed in this chapter can be applied to solving large scale portfolio selection problems.

This chapter is organized as follows. We present the mean-variance-skewness model in Section 7.2 and an approximate model in Section 7.3. A numerical example is given in Section 7.4 to illustrate the method. Some conclusions are finally given in Section 7.5.

7.2 Mean-Variance-Skewness Model

We consider a capital market with n risky assets offering random rates of returns and a risk-less asset offering a fixed rate of return. An investor allocates his/her wealth among the n risky assets and the risk-less asset to pursue the maximization of the expected utility of his/her end-of period wealth. First we make a few notations as follows:

x_i : the proportion invested in risky asset $i, i = 1,\dots,n$;

x_{n+1} : the proportion invested in the risk-less asset;

r_i : the random rate of return on the risky asset $i, i = 1,\dots,n$;

R_{n+1} : the rate of return on risk-less asset;

R_i : $E(r_i)$, the expected rate of return on risky asset $i, i = 1,\dots,n$;

σ_{ij} : $\mathrm{cov}(r_i, r_j)$, the covariance between r_i and r_j, $i, j = 1,\dots,n$;

γ_{ijk} : $E\lfloor(r_i - R_i)(r_j - R_j)(r_k - R_k)\rfloor$, the central third moment of returns;

c_i : the transaction cost of the i -th risky asset, $i = 1,\dots,n$;

k_i : the constant cost per change in a proportion of the i -th risky asset, $k_i \geq 0, i = 1,\dots,n.$

A portfolio $x = (x_1,\dots,x_n,x_{n+1})$ can result in the random rate of return:

$$r(x) = \sum_{i=1}^{n} r_i x_i + R_{n+1} x_{n+1}$$

Same as in Markowitz (1987), Yoshimoto (1996), Perold (1984) and Li, Wang and Deng (2000), it is assumed that the transaction cost c_i is a V-shaped function of the difference between an existing portfolio $x^0 = (x_1^0,\dots,x_n^0,x_{n+1}^0)$ and a new portfolio $x = (x_1,\dots,x_n,x_{n+1})$, and is formulated explicitly into the portfolio return. Thus, the transaction cost of risky asset i can be expressed as $c_i = k_i \left| x_i - x_i^0 \right|, i = 1,\dots,n$. The total transaction cost is $\sum_{i=1}^{n} c_i = \sum_{i=1}^{n} k_i \left| x_i - x_i^0 \right|$, and the net expected return, the variance and the skewness of portfolio $x=(x_1,\ldots,x_n,x_{n+1})$ are

$$R(x)=\sum_{i=1}^{n+1}R_ix_i-\sum_{i=1}^{n}k_i\left|x_i-x_i^0\right|$$

$$V(x)=\sum_{i=1}^{n}\sum_{j=1}^{n}\sigma_{ij}x_ix_j$$

and

$$S(x)=\sum_{i=1}^{n}\sum_{j=1}^{n}\sum_{k=1}^{n}\gamma_{ijk}x_ix_jx_k$$

respectively. For a new investor, we can set $x_i^0=0, i=1,\ldots,n$.

Assume that the investor is a risk-averse individual who displays decreasing absolute risk aversion. Let $u(x)$ be the utility function of an investor, then from the description of the individual, the first three derivatives satisfy that $u'(x)\geq 0$, $u''(x)\leq 0$, $u'''(x)\geq 0$. The investor's goal is to select an optimal portfolio x to maximize his/her expected value of $u[r(x)]$. Thus, the problem can be mathematically stated as (P_1):

$$\begin{aligned} &\text{maximize} \quad E[u(r(x))] \\ &\text{subject to} \quad \textstyle\sum_{i=1}^{n+1}x_i=1 \\ &\qquad\qquad x_i\geq 0, \quad i=1,\ldots,n+1 \end{aligned}$$

The constraints imply that a fund will be fully invested and that short sales and borrowings are not allowed.

Assume that $u[r(x)]$ can be approximated by the third order Taylor's expansion around the mean $E(r(x))$ of $r(x)$:

$$u[r(x)]=u[E(r(x))]+u'[E(r(x))]\,[r(x)-E(r(x))]+$$
$$\frac{1}{2}u''[E(r(x))]\,[r(x)-E(r(x))]^2+\frac{1}{6}u'''[E(r(x))]\,[r(x)-E(r(x))]^3$$

Hence, we have

$$E[u(r(x))]=u[E(r(x))]+\frac{1}{2}u''[E(r(x))]\,E\,[r(x)-E(r(x))]^2$$
$$+\frac{1}{6}u'''[E(r(x))]\,E\,[r(x)-E(r(x))]^3$$

If the distribution of $r(x)$ is symmetric around $E(r(x))$, then the expected utility function can be approximated by a function of the mean and the variance of the rate of the return of the portfolio because the third term vanishes in this case. Therefore, it is reduced to the mean-variance model in this case.

The mean-variance-skewness model proposed here is a three-objective programming problem. The optimal selection should maximize both the expected return and the skewness as well as minimize the variance. The problem can be stated as (P_2):

$$\text{maximize } R(x)$$

$$\text{minimize } V(x)$$

$$\text{maximize } S(x)$$

$$\text{subject to } \sum_{i=1}^{n+1} x_i = 1,$$

$$x_i \geq 0, \quad i = 1, \ldots, n+1$$

A portfolio $x = (x_1, \ldots, x_{n+1})$ is said to be feasible to (P_2) if it satisfies all the constraints of (P_2). A feasible portfolio x^* is said to be efficient if there exists no other feasible portfolio x such that $R(x) \geq R(x^*), V(x) \leq V(x^*)$ and $S(x) \geq S(x^*)$ with at least one strict inequality. The set of all the efficient portfolios is called the efficient frontier of (P_2). An efficient portfolio is one that maximizes $S(x)$ under given levels of expected return $R(x)$ and variance $V(x)$. According to the theory of multi-objective optimization, an efficient portfolio can be found by solving the following parametric programming problem (P_3):

$$\text{maximize } \mathrm{S(x)} = \sum_{i=1}^{n} \sum_{j=1}^{n} \sum_{k=1}^{n} \gamma_{ijk} x_i x_j x_k$$

$$\text{subject to } R(x) = \sum_{i=1}^{n+1} R_i x_i - \sum_{i=1}^{n} k_i \left| x_i - x_i^0 \right| \geq l$$

$$V(x) = \sum_{i=1}^{n} \sum_{j=1}^{n} \sigma_{ij} x_i x_j \leq \sigma^2$$

$$\sum_{i=1}^{n+1} x_i = 1,$$

$$x_i \geq 0, \quad i = 1, \ldots, n+1$$

where l, σ are appropriate parameters. Problem (P_3) is not easy to be solved because it is a non-convex and non-smooth optimization problem. In the next section, we will discuss how to convert this non-convex and non-smooth programming problem to a relatively simple problem to find an efficient solution to (P_2).

7.3 An Approximate Model

The key difficulty in solving (P_3) is the non-convex and non-smooth term in the constraint of the problem. The absolute deviation rather than the variance taken as the measure of risk in portfolio selection was analyzed by a few authors (e.g., [Konno (1990), Konno and Yamazaki (1991), Konno *et al.* (1993)]).

Let r be a random variable. The absolute deviation of r is defined as

$$W(r) = E|r - E(r)|$$

We quote a result of Konno (1990) as follows.

Theorem 7.1. If $(r_1,\dots,r_n)$ is multivariate normally distributed, then

$$W(\sum_{i=1}^{n} r_i x_i) = \sqrt{\frac{2}{\pi} \quad V(\sum_{i=1}^{n} r_i x_i)} \ .$$

The absolute deviation is equivalent to the standard deviation under a normality assumption. There is a very high correlation between them even without the normality assumption [Konno *et al.* (1993)]. Thus, we can replace a quadratic inequality constraint by a piecewise linear one.

The second difficulty in solving the problem relates the term "skewness" which can be also approximated by a piecewise linear function. Let $f(x) = x^3$. The skewness can be written as

$$S(x) = E[f(\sum_{i=1}^{n} x_i r_i - E(\sum_{i=1}^{n} x_i r_i))]$$

Let $g(x)$ be a piecewise linear function which can approximate of $f(x) = x^3$ locally. If the points $\{a_i\}$ are given as

$$a_{-(k+1)} < a_{-k} < \dots < a_{-2} < a_{-1} < 0 = a_0 < a_1 \text{, where } a_{-1} = -a_1$$

We can construct $g(x)$ by the following analytical form:

$$g(x) = \begin{cases} a_i^3 + \dfrac{a_i^3 - a_{i+1}^3}{a_i - a_{i+1}}(x - a_i) & \text{if } a_i \le x \le a_{i+1}, i = -k,\dots,0 \\ a_{-k}^3 + \dfrac{a_{-(k+1)}^3 - a_{-k}}{a_{-(k+1)} - a_{-k}}(x - a_{-k}) & \text{if } x \le a_{-k} \\ a_1^2 x & \text{if } x \ge a_1 \end{cases}$$

Instead of maximizing the third moment, we can maximize the expected value of function $g(x)$. Hence, we have the maximization problem (P_4):

$$\begin{aligned} &\text{maximize} \quad E[g(\textstyle\sum_{i=1}^{n} x_i r_i - \sum_{i=1}^{n} x_i R_i)] \\ &\text{subject to} \quad \textstyle\sum_{i=1}^{n+1} x_i R_i - \sum_{i=1}^{n} k_i \left| x_i - x_i^0 \right| \geq l \\ &\qquad E\left| \textstyle\sum_{i=1}^{n} x_i r_i - \sum_{i=1}^{n} x_i R_i \right| \leq \omega \\ &\qquad \textstyle\sum_{i=1}^{n+1} x_i = 1, \\ &\qquad x_i \geq 0, \quad i = 1,\ldots,n+1 \end{aligned}$$

By the definition of $g(x)$, we observe that if x varies along the segment $[a_{-(k+1)}, a_1]$, then $g(x)$ can be regarded as a good local approximation of $f(x)$ if the points are properly selected. When x varies in the right side of the point a_1, then $g(x)$ lies below $f(x)$. Let x be an optimal solution of problem (P_4). Then the portfolio x is expected to have a shorter tail to the left of the mean. Hence, it is expected to have a relatively large third moment if there is not a maximal one. This method meets the requirement of an investor to have smaller change of big loss below the required average rate of return $r(x)$.

$g(x)$ is a concave function and can be written as the following equivalent form:

$$g(x) = g_0(x) + g_{-1}(x) + \ldots + g_{-k}(x)$$

where

$$g_0(x) = a_1^2 x$$

and

$$g_i(x) = \begin{cases} \left(\dfrac{a_i^3 - a_{i-1}^3}{a_i - a_{i-1}} - \dfrac{a_i^3 - a_{i+1}^3}{a_i - a_{i+1}}\right)(x - a_i), \text{if } x \leq a_i, i = -1,\ldots,-k \\ 0, \qquad\qquad \text{if } x > a_i \end{cases}$$

If we know the realization $r_{it} (i = 1,\ldots,n; t = 1,\ldots,T)$ of r_i at period t, then the sampling moments can be incorporated to the model and a very useful representation of (P_4) is the following problem (P_5):

$$\begin{aligned} &\text{maximize} \quad O_1(x) = \tfrac{1}{T-1} \textstyle\sum_{t=1}^{T} \sum_{j=-k}^{0} g_j(\sum_{i=1}^{n} x_i r_{it} - \sum_{i=1}^{n} x_i R_i) \\ &\text{subject to} \quad \textstyle\sum_{i=1}^{n+1} x_i R_i - \sum_{i=1}^{n} k_i \left| x_i - x_i^0 \right| \geq l \end{aligned}$$

$$\frac{1}{T-1}\sum_{t=1}^{T}\left|\sum_{i=1}^{n} x_i r_{it} - \sum_{i=1}^{n} x_i R_i\right| \le \omega$$

$$\sum_{i=1}^{n+1} x_i = 1,$$

$$x_i \ge 0, \quad i = 1,\ldots,n+1$$

where $R_i = \frac{1}{T}\sum_{t=1}^{T} r_{it}$, $i = 1,\ldots,n$.

Now we discuss how to transform (P_5) into a linear programming problem. Problem (P_5) can be equivalently formulated as (P_6):

$$\text{maximize } \; O_1(x) = \frac{1}{T-1}\sum_{t=1}^{T}\sum_{j=-K}^{0} g_j\left(\sum_{i=1}^{n} x_i r_{it} - \sum_{i=1}^{n} x_i R_i\right)$$

$$\text{subject to } \; \sum_{i=1}^{n} x_i R_i - \sum_{i=1}^{n} K_i(c_i^+ + c_i^-) \ge l$$

$$c_i^+ - c_i = x_i - x_i^0, \; i = 1,\ldots,n$$

$$\frac{1}{T-1}\sum_{t=1}^{T}(d_t^+ + d_t^-) \le w, d_t^+ - d_t^- = \sum_{i=1}^{n} x_i r_{it} - \sum_{i=1}^{n} x_i R_i$$

$$c_i^+ \ge 0, \; i = 1,\ldots,n$$

$$c_i^- \ge 0, \; i = 1,\ldots,n$$

$$c_i^+ c_i^- = 0, \quad i = 1,\ldots,n$$

$$d_t^+ \ge 0, \quad t = 1,\ldots,T$$

$$d_t^- \ge 0, \quad t = 1,\ldots,T$$

$$d_t^+ d_t^- = 0, \; t = 1,\ldots,T$$

$$\sum_{i=1}^{n+1} x_i = 1$$

$$x_i \ge 0, \; i = 1,\ldots, n$$

where $R_i = \frac{1}{T}\sum_{t=1}^{T} r_{it}$, $i = 1,\ldots,n$.

Theorem 7.2. $(x_1^*,\ldots,x_{n+1}^*)$ is an optimal solution of (P_5) if and only if there exist $c_1^{+*},\ldots,c_n^{+*}, c_1^{-*},\ldots,c_n^{-*}, d_1^{+*},\ldots,d_T^{+*}, d_1^{-*},\ldots,d_T^{-*}$ such that ($x_1^*,\ldots,x_{n+1}^*, c_1^{+*},\ldots,c_n^{+*}$, $c_1^{-*},\ldots,c_n^{-*}, d_1^{+*},\ldots,d_T^{+*}, d_1^{-*},\ldots,d_T^{-*}$) is an optimal solution of (P_6).

Proof. Assume that $(x_1,\ldots,x_{n+1})$ is an optimal solution of (P_5). Let

$$c_i^+ = \frac{1}{2}\left|x_i - x_i^0\right| + \frac{1}{2}(x_i - x_i^0),$$

$$c_i^- = \frac{1}{2}\left|x_i - x_i^0\right| - \frac{1}{2}(x_i - x_i^0)$$

$$d_t^+ = \frac{1}{2}\left|\sum_{i=1}^n x_i r_{it} - \sum_{i=1}^n x_i R_r\right| + \frac{1}{2}(\sum_{i=1}^n x_i r_{it} - \sum_{i=1}^n x_i R_r)$$

$$d_t^- = \frac{1}{2}\left|\sum_{i=1}^n x_i r_{it} - \sum_{i=1}^n x_i R_r\right| - \frac{1}{2}(\sum_{i=1}^n x_i r_{it} - \sum_{i=1}^n x_i R_r)$$

Thus, equivalent expressions are

$$c_i^+ + c_i^- = \left|x_i - x_i^0\right|,$$

$$c_i^+ - c_i^- = x_i - x_i^0,$$

$$c_i^+ c_i^- = 0, c_i^+ \geq 0, c_i^- \geq 0$$

$$d_t^+ + d_t^- = \left|\sum_{i=1}^n x_i r_{it} - \sum_{i=1}^n x_i R_i\right|, d_t^+ - d_t^- = \sum_{i=1}^n x_i r_{it} - \sum_{i=1}^n x_i R_i$$

$$d_t^+ d_t^- = 0, d_t^+ \geq 0, d_t^- \geq 0$$

Hence, ($x_1,...,x_{n+1}$, $c_1^+,...,c_n^+, c_1^-,...,c_n^-, d_1^+,...,d_T^+, d_1^-,...,d_T^-$) is a feasible solution of (P_6).

Conversely, let ($x_1^*,...,x_{n+1}^*$, $c_1^{+*},...,c_n^{+*}, c_1^{-*},...,c_n^{-*}, d_1^{+*},...,d_T^{+*}, d_1^{-*},...,d_T^{-*}$) be an optimal solution of (P_6). It is easy to verify that $(x_1^*,...,x_{n+1}^*)$ is a feasible solution of (P_5).

Because the objective functions of (P_5) and (P_6) are the same, we have

$$O_1(x_1,...,x_{n+1}) \leq O_1(x_1^*,...,x_{n+1}^*) \leq O_1(x_1,...,x_{n+1}).$$

Thus,

$$O_1(x_1,...,x_{n+1}) = O_1(x_1^*,...,x_{n+1}^*)$$

and ($x_1,...,x_{n+1}$, $c_1^+,...,c_n^+, c_1^-,...,c_n^-, d_1^+,...,d_T^+, d_1^-,...,d_T^-$) is an optimal solution of (P_6). Therefore, $(x_1^*,...,x_{n+1}^*)$ is an optimal solution of (P_5). This completes the proof.

Eliminating the constrains $c_i^+ c_i^- = 0, d_t^+ d_t^- = 0, i = 1,...,n, t = 1,...,T$ in (P_6), we get the following problem (P_7):

$$\text{maximize } \ O_1(x)=\frac{1}{T-1}\sum_{t=1}^{T}\sum_{j=-K}^{0} g_j(\sum_{i=1}^{n} x_i r_{it}-\sum_{i=1}^{n} x_i R_i)$$

$$\text{subject to } \ \sum_{i=1}^{n} x_i R_i - \sum_{i=1}^{n} K_i(c_i^+ + c_i^-)\geq l,$$

$$c_i^+ - c_i^- = x_i - x_i^0, \ \ i=1,\dots,n$$

$$\frac{1}{T-1}\sum_{t=1}^{T}(d_t^+ + d_t^-)\leq w, d_t^+ - d_t^- = \sum_{i=1}^{n} x_i r_{it}-\sum_{i=1}^{n} x_i R_i$$

$$c_i^+\geq 0, \ \ i=1,\dots,n$$

$$c_i^-\geq 0, \ \ i=1,\dots,n$$

$$d_t^+\geq 0, \ \ t=1,\dots,T$$

$$d_t^-\geq 0, \ \ t=1,\dots,T$$

$$\sum_{i=1}^{n+1} x_i = 1$$

$$x_i\geq 0, i=1,\dots,n$$

We can prove the following theorem.

Theorem 7.3. Let ($x_1^*,\dots,x_{n+1}^*, c_1^{+*},\dots,c_n^{+*}, c_1^{-*},\dots,c_n^{-*}, d_1^{+*},\dots,d_T^{+*}, d_1^{-*},\dots,d_T^{-*}$) be an optimal solution of (P_7). Then there exist $\tilde{c}_1^+,\dots,\tilde{c}_n^+,\tilde{c}_1^-,\dots,\tilde{c}_n^-,\tilde{d}_1^+,\dots,\tilde{d}_T^+,\tilde{d}_1^-,\dots,\tilde{d}_T^-$, such that ($x_1^*,\dots,x_{n+1}^*, \tilde{c}_1^+,\dots,\tilde{c}_n^+,\tilde{c}_1^-,\dots,\tilde{c}_n^-,\tilde{d}_1^+,\dots,\tilde{d}_T^+,\tilde{d}_1^-,\dots,\tilde{d}_T^-$) is an optimal solution of (P_6).

Proof. Assume that ($x_1^*,\dots,x_{n+1}^*, c_1^{+*},\dots,c_n^{+*}, c_1^{-*},\dots,c_n^{-*}, d_1^{+*},\dots,d_T^{+*}, d_1^{-*},\dots,d_T^{-*}$) is an optimal solution of (P_7). Let

$$\tilde{c}_i^+ = \begin{cases} c_i^{+*} - c_i^{-*}, \text{if } c_i^{+*} > c_i^{-*} > 0 \\ 0, \quad \text{if } c_i^{-*} \geq c_i^{+*} > 0 \end{cases}$$

$$\tilde{c}_i^- = \begin{cases} 0, \quad \text{if } c_i^{+*} > c_i^{-*} > 0 \\ c_i^{-*} - c_i^{+*}, \text{if } c_i^{-*} \geq c_i^{+*} > 0 \end{cases}$$

$$\begin{pmatrix} \tilde{c}_i^+ \\ \tilde{c}_i^- \end{pmatrix} = \begin{pmatrix} c_i^{+*} \\ c_i^{-*} \end{pmatrix}, \text{if } c_i^{+*} c_i^{-*} = 0$$

$$\tilde{d}_j^+ = \begin{cases} d_j^{+*} - d_j^{-*}, \text{if } d_j^{+*} > d_j^{-*} > 0 \\ 0, \quad \text{if } d_j^{-*} \geq d_j^{+*} > 0 \end{cases},$$

$$\tilde{d}_j^- = \begin{cases} 0, & \text{if } d_j^{+*} > d_j^{-*} > 0 \\ d_j^{-*} - d_j^{+*}, & \text{if } d_j^{-*} \geq d_j^{+*} > 0 \end{cases}$$

$$\begin{pmatrix} \tilde{d}_j^+ \\ \tilde{d}_j^- \end{pmatrix} = \begin{pmatrix} d_j^{+*} \\ d_j^{-*} \end{pmatrix}, \text{if } \mathrm{d}_j^{+*} d_j^{-*} = 0$$

Then ($x_1^*,\ldots,x_{n+1}^*$, $\tilde{c}_1^+,\ldots,\tilde{c}_n^+,\tilde{c}_1^-,\ldots,\tilde{c}_n^-,\tilde{d}_1^+,\ldots,\tilde{d}_T^+,\tilde{d}_1^-,\ldots,\tilde{d}_T^-$) is a feasible solution of (P_6). It is obvious that each feasible solution of (P_6) is feasible to (P_7). Because the objective functions of (P_6) and (P_7) are the same, ($x_1^*,\ldots,x_{n+1}^*$, $\tilde{c}_1^+,\ldots,\tilde{c}_n^+,\tilde{c}_1^-,\ldots,\tilde{c}_n^-$, $\tilde{d}_1^+,\ldots,\tilde{d}_T^+,\tilde{d}_1^-,\ldots,\tilde{d}_T^-$) is optimal to (P_6). The theorem is proved.

Because the objective function of problem (P_7) is piecewise linear, we can replace it by a linear objective function so that the problem can be correspondingly transformed to a linear programming problem. Using a similar technique as above, we have an equivalent optimization problem (P_8):

$$\text{maximize } \; \mathrm{O}_2(x) = \tfrac{1}{T-1}\sum_{t=1}^{T}\sum_{j=-k}^{-1}\left(\frac{a_j^3 - a_{j-1}^3}{a_j - a_{j-1}} - \frac{a_j^3 - a_{j+1}^3}{a_j - a_{j+1}}\right)u_{jt}$$

$$\text{subject to } \; \sum_{i=1}^{n+1} x_i R_i - \sum_{i=1}^{n} k_i (c_i^+ + c_i^-) \geq l$$

$$c_i^+ \geq 0, \; i = 1,\ldots,n+1$$

$$c_i^- \geq 0, \; i = 1,\ldots,n+1$$

$$c_i^+ - c_i^- = x_i - x_i^0, \; i = 1,\ldots,n+1$$

$$\tfrac{1}{T-1}\sum_{t=1}^{T}(d_t^+ + d_t^-) \leq \omega$$

$$d_t^+ \geq 0, \; t = 1,\ldots,T$$

$$d_t^- \geq 0, \; t = 1,\ldots,T$$

$$d_t^+ - d_t^- = \sum_{i=1}^{n} x_i r_{it} - \sum_{i=1}^{n} x_i R_i, \; t = 1,\ldots,T$$

$$u_{jt} \leq 0, \; t = 1,\ldots,T, \; j = -K,\ldots,-1$$

$$u_{jt} - \left(\sum_{i=1}^{n} x_i r_{it} - \sum_{i=1}^{n} x_i R_i - a_j\right) \leq 0, \; t = 1,\ldots,T, \; j = -K,\ldots,-1$$

$$\sum_{i=1}^{n+1} x_i = 1$$

$$x_i \geq 0, \quad i = 1,\dots,n+1$$

Theorem 7.4. ($x_1^*,\dots,x_{n+1}^*$, $\widetilde{c}_1^+,\dots,\widetilde{c}_n^+,\widetilde{c}_1^-,\dots,\widetilde{c}_n^-,\widetilde{d}_1^+,\dots,\widetilde{d}_T^+,\widetilde{d}_1^-,\dots,\widetilde{d}_T^-$) is an optimal solution of problem (P_7) if and only if there exist $u_{jt}(j=-K,\dots,-1; t=1,\dots,T)$ such that $(x_1^*,\dots,x_{n+1}^*,\widetilde{c}_1^+,\dots,\widetilde{c}_n^+,\widetilde{c}_1^-,\dots,\widetilde{c}_n^-,\widetilde{d}_1^+,\dots,\widetilde{d}_T^+,\widetilde{d}_1^-,\dots,\widetilde{d}_T^-,u)$ is an optimal solution of (P_8).

Proof. Assume that $(x_1,\dots,x_{n+1},c_1^+,\dots,c_n^+,c_1^-,\dots,c_n^-,d_1^+,\dots,d_T^+,d_1^-,\dots,d_T^-)$ is an optimal solution of problem (P_7). Let

$$u_{jt} = \begin{cases} \sum_{i=1}^{n+1} x_i r_{it} - \sum_{i=1}^{n+1} x_i R_i - a_j, & \text{if } \sum_{i=1}^{n+1} x_i r_{it} - \sum_{i=1}^{n+1} x_i R_i \leq a_j, \ \text{j = -K,...,-1, t = 1,...,T} \\ 0, & \text{if } \sum_{i=1}^{n+1} x_i r_{it} - \sum_{i=1}^{n+1} x_i R_i > a_j \end{cases}$$

Then $(x_1,\dots,x_{n+1},c_1^+,\dots,c_n^+,c_1^-,\dots,c_n^-,d_1^+,\dots,d_T^+,d_1^-,\dots,d_T^-,u)$ is a feasible solution of (P_8) and $O_1(x_1,\dots,x_{n+1}) = O_2(x_1,\dots,x_{n+1},u)$.

Assume that $(x_1^*,\dots,x_{n+1}^*,\widetilde{c}_1^+,\dots,\widetilde{c}_n^+,\widetilde{c}_1^-,\dots,\widetilde{c}_n^-,\widetilde{d}_1^+,\dots,\widetilde{d}_T^+,\widetilde{d}_1^-,\dots,\widetilde{d}_T^-,u)$ is an optimal solution of (P_8). It is not hard to show that ($x_1^*,\dots,x_{n+1}^*$, $\widetilde{c}_1^+,\dots,\widetilde{c}_n^+,\widetilde{c}_1^-,\dots,\widetilde{c}_n^-,\widetilde{d}_1^+,\dots,\widetilde{d}_T^+,\widetilde{d}_1^-,\dots,\widetilde{d}_T^-$) is a feasible solution of problem (P_7).

Because

$$\frac{a_j^3 - a_{j-1}^3}{a_j - a_{j-1}} - \frac{a_j^3 - a_{j+1}^3}{a_j - a_{j+1}} = (a_{j-1} - a_{j+1})(a_{j-1} + a_j + a_{j+1}) > 0, j=-k,\dots,-1$$

$$a_{j-1} < a_j < a_{j+1} < 0, j = -K,\dots,-1$$

and by the optimality, $\widetilde{u}_{jt}$ must satisfy

$$\widetilde{u}_{jt} = \begin{cases} \sum_{i=1}^{n+1} x_i^* r_{it} - \sum_{i=1}^{n+1} x_i^* R_i - a_j, \text{if } \sum_{i=1}^{n+1} x_i^* r_{it} - \sum_{i=1}^{n+1} x_i^* R_i \leq a_j \\ 0, \quad \text{if } \sum_{i=1}^{n+1} x_i^* r_{it} - \sum_{i=1}^{n+1} x_i^* R_i > a_j \end{cases}$$

j=-K,...,-1; t=1,...,T. Thus, $O_1(x_1^*,\dots,x_{n+1}^*) = O_2(x_1^*,\dots,x_{n+1}^*,\widetilde{u})$.

Hence,

$$\begin{aligned} O_1(x_1,\dots,x_{n+1}) &= O_2(x_1,\dots,x_{n+1},u) \leq O_2(x_1^*,\dots,x_{n+1}^*,\widetilde{u}) \\ &= O_1(x_1^*,\dots,x_{n+1}^*) \leq O_1(x_1,\dots,x_{n+1}) \end{aligned}$$

Therefore, $O_1(x_1,...,x_{n+1}) = O_2(x_1,...,x_{n+1},u) = O_2(x_1^*,...,x_{n+1}^*,\tilde{u}) = O_1(x_1^*,...,x_{n+1}^*)$
Consequently, $(x_1,...,x_{n+1},c_1^+,...,c_n^+,c_1^-,...,c_n^-,d_1^+,...,d_T^+,d_1^-,...,d_T^-,u)$ is optimal to (P_8) and $(x_1^*,...,x_{n+1}^*,\tilde{c}_1^+,...,\tilde{c}_n^+,\tilde{c}_1^-,...,\tilde{c}_n^-,\tilde{d}_1^+,...,\tilde{d}_T^+,\tilde{d}_1^-,...,\tilde{d}_T^-)$ is an optimal solution of problem (P_7). The theorem is proved.

By Theorems *3, 4* and *5*, we can solve the general mean-variance-skewness model (P_3) approximately by solving the linear programming problem (P_8). Problem (P_8) is with $(K+2)T+3n+1$ variables and $(K+1)T+n+3$ constrains and can be easily solved with a few numerical methods such as a simplex method.

7.4 A Numerical Example

In this section, we give a numerical example to illustrate the method for portfolio selection proposed in this chapter.

Consider the following portfolio selection problem. The returns of six stocks at eight periods are given in Table 7.1.

Table 7.1 Returns of six stocks during eight periods

Period	Stock 1	Stock 2	Stock 3	Stock4	Stock5	Stock 6
1	0.03	0.14	0.13	-0.11	0.05	0.01
2	-0.01	0.03	0.12	0.07	-0.07	0.08
3	0.09	-0.05	0.05	0.14	0.03	-0.06
4	0.24	-0.07	-0.07	0.21	0.09	0.07
5	-0.09	0.05	0.08	0.04	0.01	0.14
6	-0.02	0.11	0.04	0.07	-0.08	-0.10
7	0.05	0.05	-0.04	0.03	0.04	0.13
8	0.11	0.04	0.03	-0.06	0.06	-0.06
mean	0.05	0.0375	0.0425	0.0487	0.0163	0.0263

The return of the risk-less asset is R_7 =0.03. The existing portfolio is x^0=(0.1,0.2,0.1,0.2,0.1,0.2,0.1) and the constant cost per change in a proportion of the risk assets is (0.003,0.001,0.005,0.004,0.003,0.002).

With the information of the historical data r_{it} and the return of the given risk-less asset R_{n+1}, we select the best portfolio with the following procedures:

a. Compute $R_i = \frac{1}{T}\sum_{t=1}^{T} r_{it}$ and find a possible upper level of l by solving

$$
\begin{aligned}
\text{maximize } & \sum_{i=1}^{7} x_i R_i - \sum_{i=1}^{6} k_i (c_i^+ + c_i^-) \\
\text{subject to } & \sum_{i=1}^{7} x_i = 1 \\
& x_i \geq 0,\ i = 1,\ldots,7 \\
& c_i^+ - c_i^- = x_i - x_i^0,\ i = 1,\ldots,6 \\
& c_i^+ \geq 0,\ i = 1,\ldots,6 \\
& c_i^- \geq 0,\ i = 1,\ldots,6
\end{aligned}
$$

We get l =0.0463.

b. For each possible level l, compute the minimal value of the absolute-deviation $\tilde{\omega}$ by solving

$$
\begin{aligned}
\text{minimize } & \tilde{w} = \sum_{t=1}^{T} (d_t^+ + d_t^-) \\
\text{subject to } & \sum_{i=1}^{7} x_i = 1 \\
& x_i \geq 0,\ i = 1,\ldots,7 \\
& c_i^+ - c_i^- = x_i - x_i^0,\ i = 1,\ldots,6 \\
& c_i^+ \geq 0,\ i = 1,\ldots,6 \\
& c_i^- \geq 0,\ i = 1,\ldots,6 \\
& d_t^+ \geq 0,\ t = 1,\ldots,T \\
& d_t^- \geq 0,\ t = 1,\ldots,T \\
& d_t^+ - d_t^- = \sum_{i=1}^{6} x_i r_{it} - \sum_{i=1}^{6} x_i R_i, \quad t = 1,\ldots,T \\
& \sum_{i=1}^{7} x_i R_i - \sum_{i=1}^{6} k_i (c_i^+ + c_i^-) \geq l
\end{aligned}
$$

We determine the possible risk tolerant level $w = \alpha\tilde{w}$, where $\alpha > 1$ is a parameter determined by the investor. The results are listed in Table 7.2.

Table 7.2 Risk tolerant level ($w = \alpha\tilde{w}, \alpha = 1.1$)

l_i	0.046	0.044	0.042	0.040
$\tilde{w}_i$	0.4036	0.1905	0.0893	0.0706
w_i	0.4439	0.2096	0.0983	0.0777

c. Estimate the lowest possible value of x (*i.e.*, $a_{-(K+1)}$) in $g(x)$ and determine points $\{a_{-(K+1)},...,a_0\}$ and $a = \min\{r_{it} - R_i, i = 1,...,n; t = 1,...,T\}$.

We get $a = \min\{r_{it} - R_i, i = 1,...,n; t = 1,...,T\}$=-0.1587. We select

$$a_{-5} = -0.15, a_{-4} = -0.12, a_{-3} = -0.09, a_{-2} = -0.06, a_{-1} = -0.03 = -a_1, a_0 = 0$$

d. Solve Problem (P_8) to get an optimal portfolio x^*.

With different coefficients, we get the corresponding results of x^* listed in Tables 7.*3*, 7.*4*, 7.*5* and 7.*6* as follows.

Table 7.3 Investment proportion with coefficients of $l_1 = 0.046, w_1 = 0.4439$

0.1726	0.144	0.1179	0.2292	0.0747	0.2039	0.0579

Table 7.4 Investment proportion with coefficients of $l_1 = 0.044, w_1 = 0.2096$

0.1667	0.0987	0.1268	0.2373	0.0711	0.1851	0.1143

Table 7.5 Investment proportion with coefficients of $l_1 = 0.042, w_1 = 0.0983$

0.0016	0.0238	0.1	0.1292	0.0006	0.0843	0.5905

Table 7.6 Investment proportion with coefficients of $l_1 = 0.040, w_1 = 0.0777$

0.0169	0.0324	0	0.2037	0	0.0328	0.7142

7.5 Conclusions

A mean-variance-skewness model has been proposed in this chapter for portfolio selection with transaction costs. With a transformation, we converted the nonconvex and nonsmooth programming problem into a linear programming problem. This implies that we can efficiently solve large-scale investment problems for mutual funds or other financial institutions in the world wide financial markets with a linear programming technique such as a simplex method.

8. Capital Asset Pricing: Theory and Methodologies

8.1 Introduction

Investors prefer returns and dislike risks. They pursue the highest risk premia (the difference between the expected returns and the riskless interest rate) in financial markets. Modern finance theory helps investors make their choice about how to allocate their limited funds to different assets. Portfolio theory which is originally proposed by Markowitz (1959) is considered the beginning of modern finance theory. Since then, a lot of researchers came up with many models and algorithms during the last five decades [Deng, Wang and Xia (2000)].

Building on the work of Markowitz, Sharpe (1964) proposed the capital asset pricing model (CAPM) in 1964. The CAPM is considered a centerpiece of modern financial economics. And since then, it is a widely used model for asset pricing in practice. It provides a precise prediction of the relationship between the risk of an asset and its expected return. Not only does this relationship provide a benchmark rate of return for evaluating possible investment, but also helps making good guess as to the expected return on assets that have not yet been traded in the marketplace.

The exploitation of security mis-pricing to earn riskfree economic profits is called arbitrage. The Arbitrage Pricing Theory (APT) developed by Ross (1976) uses a no-arbitrage argument to derive the model. No-arbitrage means that you could not make profits by simultaneous purchase and sale of equivalent securities (often in different markets). The most basic principle of capital market theory is that equilibrium market prices should rule out arbitrage opportunities.

In this chapter, we consider the capital asset pricing model, the arbitrage pricing theory and their various derivatives. In Section 8.2, we illustrate the standard form of capital asset pricing model and in Section 8.3 we describe various non-standard forms of capital asset pricing models. The empirical tests of the CAPM are discussed in Section 8.4 and the intertemporal capital asset pricing model is explained in Section 8.5. We illustrate the consumption-based capital asset pricing model in Section 8.6 and the arbitrage pricing theory in Section 8.7. The intertemporal arbitrage pricing theory is given in Section 8.8. We presents the comparison of CAPM and APT in Section 8.9. Finally, Section 8.10 gives some conclusion remarks.

8.2 The Capital Asset Pricing Model

The capital asset pricing model, or CAPM, predicts the relationship between the risk and the equilibrium expected returns on risky assets. The model is based on some assumptions on the financial markets and assets. The assumptions are as follows:

1) No transaction costs. There is no cost (friction) of buying or selling any asset. This ignoring the transaction costs greatly reduce the complexity.

2) Infinitely divisible assets. This assumption allows the investor to take any position in an investment.

3) No personal income tax. The individual is indifferent to the form (dividend or capital gains) in which the return on the investment is received.

4) Perfect competition. An individual can not affect the price of a stock by buying or selling action. While no single investor can affect prices by an individual action, investors in total determine prices by their actions.

5) Investors are expected to make decisions solely in terms of expected values and standard deviations of the returns on their portfolios.

6) Unlimited short sales. An individual investor can sell short any amount of any shares.

7) Unlimited lending and borrowing at the riskless rate. An investor can lend or borrow any amount of funds desired at a rate of interest equal to the rate of riskless securities.

8) All investors analyze securities in the same way and share the same economic view of the world. All investors use same expected returns, standard deviations, and correlations to generate the efficient frontier and the unique optimal risky portfolio.

9) All assets are marketable. All assets, including human capital, can be sold and bought on the market.

Under the above assumptions, all investors have the same variance-covariance matrix, and every investor's efficient set will be the same. Since every investor's risky portfolio is the same, this portfolio is simply the portfolio of all the risky assets in the market. This portfolio is called the market portfolio and is depicted by the letter m. The market portfolio is a value-weighted portfolio. Each security is held in a proportion equal to its market value divided by the total market value of all the securities.

With the above assumptions, the standard Sharpe-Lintner version of CAPM is represented as:

$$E[R_i] = R_f + \beta_i(E[R_m] - R_f)$$

where $E[R_i]$ and $E[R_m]$ are respectively the expected return of stock i and the expected return of the market portfolio, R_f is the return on the riskless asset, and the $\beta_i = \sigma_{im} / \sigma_m^2$ is systematic risk measure of the i -th stock which can not be diversified through securities' combination. σ_{im} is the covariance between the return on asset i, and the return on the market portfolio and σ_m^2 is the variance of the returns on the risky assets in the market portfolio.

The above equation is called the security market line (SML) and it is shown in the following figure.

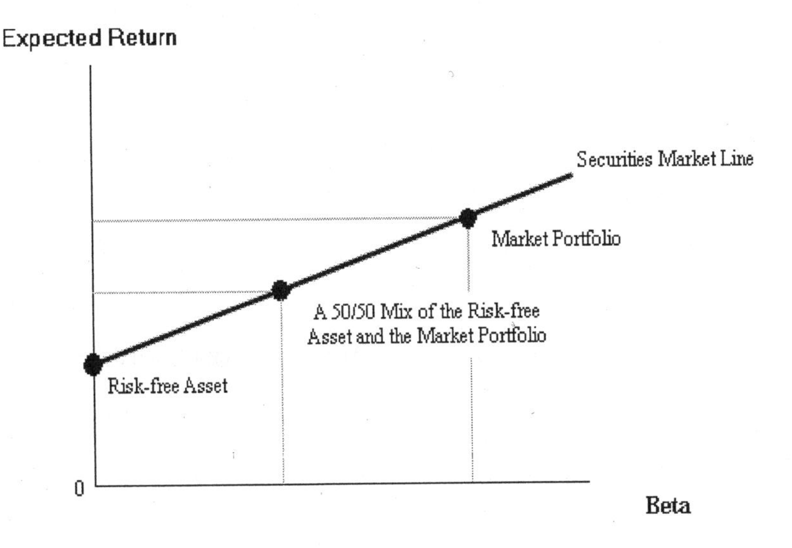

Figure 8.1 Security market line of CAPM

The characteristics of an asset Beta are similar to that of a covariance. It measures the degree to which an asset's return co-moves with the returns on the market. Because of the standardization, the Beta can be interpreted as the slope of the regression line of the asset return on the market return. This regression line is also called a characteristic line shown below.

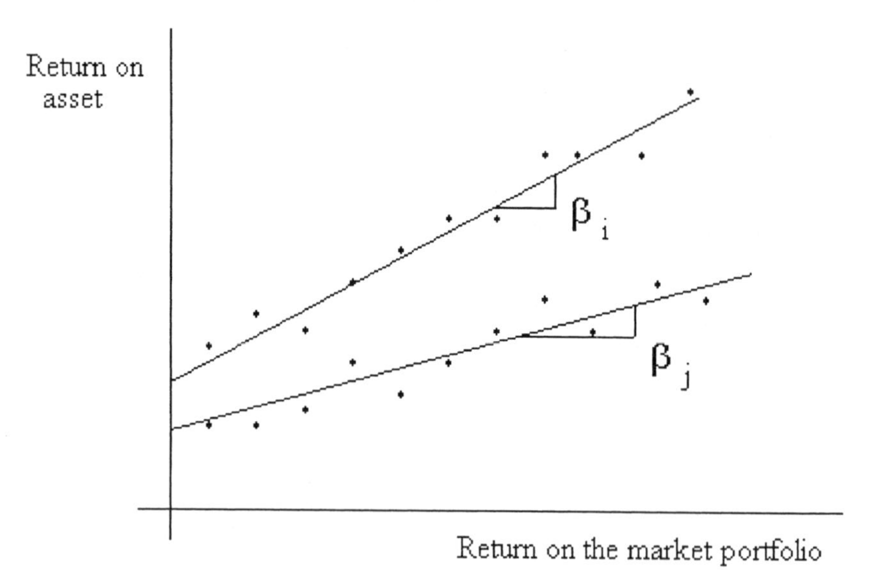

Figure 8.2 Characteristic line

In the above figure, asset i has a larger Beta than asset j. Betas also have the following properties:

1) $\beta_p = \sum w_i \beta_i$, the Beta of a portfolio is the weighted average of the Betas of the assets that make up the portfolio.

2) $\beta_m = \text{cov}(r_m, r_m)/\text{var}(r_m) = 1$, which means that the market Beta is equal to one.

If we define β_i as the price of risk, the CAPM can be explained as follows:

(Expected return)=(Price of time)+(Price of risk) × *(Amount of risk)*

At time t , the return rate of the i -th stock is

$$R_{it} = R_{ft} + \beta_i (r_{mt} - R_{ft}) + \eta_{it}$$

$$E[\eta_{it}] = 0$$

$$\text{cov}(\eta_{it}, \eta_{jt}) = 0, \quad (i \neq j)$$

$$\mathrm{cov}(R_{mt}, \eta_{it}) = 0$$

where R_{ft} and R_{mt} are respectively the returns of the riskfree asset and the market portfolio at time t.

8.3 Nonstandard Forms of Capital Asset Pricing Models

It is obvious that not all of the assumptions of the CAPM are reasonable. Also, financial markets at different countries are different with respect to the assumptions satisfied. So it is not surprising that since the publication of Sharpe's article, many authors work on the problems of relaxing the assumptions involved in the derivation of the CAPM. The model has been extended to the more general economies, but at the expense of simplicity in the structure of equilibrium expected returns. We introduce a few non-standard forms [Elton and Gruber (1995)] of CAPM as follows.

8.3.1 Modifications of riskless lending and borrowing

One assumption of the CAPM is that investors can lend and borrow unlimited sums of money at the riskless rate of interest. Such an assumption is clearly not a description of the real world. It is more realistic to assume that investors can lend unlimited sums of money at the riskless rate but cannot borrow at a riskless rate. The lending assumption is equivalent to the investors who are able to buy government securities equal in maturity to their single-period horizon. Such securities exist and the rate on such securities is virtually the same for all investors. On the other hand, it is not possible for investors to borrow unlimited amounts at a riskless rate. We first examine the case where investors can neither borrow nor lend at the riskless rate, and then to extend the analysis to the case where they can lend but not borrow at the riskless rate.

8.3.1.1 No riskless lending or borrowing

In the absence of a riskless asset, Black (1972) derived a two-factor version of the CAPM, also known as the zero-Beta CAPM. In this version, the expected return of asset i in the excess of the zero-Beta return is linearly related to its Beta. Specially, for the expected return of asset i, $E[R_i]$, we have

$$E[R_i] = E[R_{0m}] + \beta_{im}(E[R_m] - E[R_{0m}])$$

where R_m is the return on the market portfolio, and R_{0m} is the return on the zero-Beta portfolio associated with m. This portfolio is defined to be the portfolio that has the minimum variance of all portfolios uncorrelated with m. Any other uncorrelated portfolio would have the same expected return, but a higher variance. Since it is wealth in real terms that is relevant, for the Black model, returns are generally stated on an inflation-adjusted basis and β_{im} is defined in terms of real returns as follows:

$$\beta_{im} = \frac{\text{cov}[R_i, R_m]}{\text{var}[R_m]}$$

8.3.1.2 Riskless lending but no riskless borrowing

In the real world, it is reasonable to assume that investors can lend at a rate that is riskless. Individuals can place funds in government securities that have a maturity equal to their time horizon and, thus, be guaranteed of a riskless payoff at the horizon.

8.3.2 Different holding periods

Levy and Samuelson (1992) assumed that assets are traded in discrete time and that risk averse investors differ in their holding periods. They investigated the conditions under which the CAPM holds and showed that when portfolio rebalancing is allowed, the CAPM holds in four cases which are: (*a*) quadratic preferences; (*b*) one-period normal distributions when the utility is defined on the multi-period terminal wealth which is not normal; (*c*) the terminal wealth is log-normally distributed; and (*d*) the terminal wealth *W(T)* is normally distributed, but in this case diverse holding periods are not allowed. Case (*d*) is similar to the Sharpe-Lintner model with the exception that $(T-1)$ revisions are allowed.

8.3.3 Personal taxes

The simple form of the capital asset pricing model ignores the presence of taxes. However, there exist taxes on capital gains and dividends and the tax rates are different for these two forms of profit. With the differential taxes on the income and the capital gains, the following general equilibrium pricing equation for all assets and portfolios can be derived as

$$E(R_i) = R_F + \beta_i[(E(R_M) - R_F) - \tau(\delta_M - R_F)] + \tau(\delta_i - R_F)$$

where δ_M is the dividend yield (dividends divided by the price) of the market portfolio, δ_i is the dividend yield for stock i, τ is the tax factor that measures the relevant market tax on capital gains and income. τ is a function of the investors' tax rates and wealth.

The above equation reveals that a security market line is no longer sufficient to describe the equilibrium relationship. In the standard version of general equilibrium relationships the only variable associated with the individual security that affected expected return is its Beta. Now the equilibrium is described in three-dimensional spaces (R_i, β_i, δ_i) rather than the two-dimension space (R_i, β_i) and both the securities Beta and its dividend yield affect the expected return.

Wang (1995) assumed that the investor only reports a certain fraction of his/her taxable income and had the conclusion that the expected excess rates of return on the risky assets are lower than those when the tax evasion is absent. The excess return is described as follows:

$$R_j - r = \frac{1}{T}[\text{cov}(\tilde{R}_j, \tilde{C}^*) + \text{cov}(\tilde{R}_j, \textstyle\sum_k dV_k \eta_k^{'} T^k)]$$

where $T = \sum_k T^k$, $T^k = -\dfrac{U_c^k}{U_{cc}^k}$, $\eta_k^{'} = \dfrac{(h^k)^2 F_k^{''}}{(1-h^k F_k^{'})}$ for investor k, $\tilde{C}^*$ is the aggregate change in the consumption rate or $\tilde{C}^* = \sum_k dc^k$.

8.3.4 Non-marketable assets

In reality, each investor has non-marketable assets or assets that he or she will not consider marketing. If the world is divided into marketable and non-marketable assets, then a simple equation exists for the equilibrium return on all the assets. Let R_H be the one period rate of return on non-marketable assets, P_H be the total value of all non-marketable assets, and P_M be the total value of all marketable assets.

Thus, it can be shown that

$$E[R_j] = R_f + \frac{E[R_M] - Rf}{\sigma_M^2 + P_H/P_M \,\text{cov}(R_M, R_H)}[\text{cov}(R_j, R_M) + \frac{P_H}{P_M}\text{cov}(R_j, R_H)]$$

With the inclusion of non-marketable assets, the market trade-off between return and risk is different. Now, the market risk-return trade-off changes from $\dfrac{E[R_M] - R_f}{\sigma_M^2}$ to $\dfrac{E[R_M] - Rf}{\sigma_M^2 + P_H/P_M \,\text{cov}(R_M, R_H)}$.

8.3.5 Heterogeneous expectations

If investors have heterogeneous expectations, equilibrium can still be expressed in terms of expected returns, covariance, and variance, but these returns, covariance, and variances are complex weighted averages of the estimates held by different individuals. Because these weightings involve information about the utility functions of the investors, especially involving information about investors' trade-offs (marginal rate of substitution) between the expected return and the variance. But this trade-off for most utility functions is a function of wealth and, hence, prices. Thus, in general, an explicit solution to the heterogeneous expectation problem can not be reached. However, Constantinides and Duffie (1996) demonstrated, in the context of a heterogeneous agent model with a single goods and persistent income shocks, that closed form solutions for prices can be obtained through a judicious choice of the idiosyncratic income specification. Ramchand (1999) extended Constantinides and Duffie's result to a two country and two goods model with an added dimension to heterogeneity viz. heterogeneity across the countries. The equilibrium model shows that, despite the additional source of heterogeneity, closed form solutions for asset prices are obtainable so long as preferences are iso-elastic and income shocks are permanent.

Detemple and Murthy (1994) examined the behavior of the interest rate, asset prices and asset holdings in an economy with heterogeneous beliefs about the expected rate of aggregate production growth and calculated equilibrium prices for assets.

Basak (2000) presented equilibrium security price dynamics in an economy where non-fundamental risk arises from agents' heterogeneous beliefs about extraneous processes. The paper considers an economy in which two or more agents, with heterogeneous arbitrary utility functions, observe two exogenous processes: the consumption supply (aggregate endowment) process, and an extraneous process which affects none of the fundamentals. The agents have full information on the endowment process, but incomplete information on the growth. All the agents believe that the extraneous process may affect the real economic quantities and form their expectations taking this into account. Then if such an equilibrium exists, the equilibrium state price density processes of the two agents are given by

$$\xi^1(t) = U'(\varepsilon(t); {}^{1}/_{y_1}, {}^{\eta(t)}/_{y_2})$$
$$\xi^2(t) = U'(\varepsilon(t); {}^{1}/_{y_1}, {}^{\eta(t)}/_{y_2}) / \eta(t)$$

where the ratio ${}^{y_1}/_{y_2}$ satisfies the agent's budget constraint, *i.e.*,

$$E'[\int_0^T U'(\varepsilon(t); 1/y_1, y(t)/y_2) I_1(y_1 U'(\varepsilon(t); 1/y_1, y(t)/y_2)) dt]$$
$$= E'[\int_0^T U'(\varepsilon(t); 1/y_1, \eta_t/y_2) \varepsilon_1(t) dt]$$

and the stochastic weighting $\eta(t) = \xi^1(t) / \xi^2(t)$ follows

$$\frac{d\eta(t)}{\eta(t)} = -\bar{\mu}(t) dW_Z'(t)$$
$$= \bar{\mu}(t)^2 dt - \bar{\mu}(t) dW_Z^2(t)$$
$$= \bar{\mu}(t) \frac{(\mu_Z^1(t) - \mu_Z(t))}{\delta_Z(t)} dt - \bar{\mu}(t) dW_Z(t)$$

where $\delta(t),\ t \in [0, T]$ is the non-fundamental risk in the economy and W_Z is a Brownian motion.

8.3.6 Non-price-taking behavior

Central to the equilibrium-based asset pricing models is the competitive agents paradigm: each agent is atomistic relative to the market and takes prices to be unaffected by his/her actions. However, in the financial markets, one or more investors, such as mutual funds or large pension funds, can affect the market prices. The method of analysis in Lindenberg (1976, 1979) derives equilibrium conditions under all possible demands by the price affector. The price affector selects his/her portfolio to maximize utility given the equilibrium prices that will result from his or her action. Assuming that the price affector operates so as to maximize his/her utility, equilibrium conditions can then be arrived. Lindenberg found that all investors, including the price taker, hold some combination of the market portfolio and the riskless asset. However, the price affector will hold less of the riskless asset (will be less of a risk avoider) than would be the case if the price affector did not recognize the fact that his or her actions affected price. By doing so the price affector increases his/her utility. Because the price affector still holds a combination of the riskless asset and the market portfolio, the simple form of the CAPM still can be derived, but the market price of risk is lower than that it would be if all the investors were price takers.

Basak (1997) gave an asset pricing model when there exists one non-price taking agent. By retaining the usual assumptions of a complete and frictionless market and symmetric information, the equilibrium asset risk premia is given by

$$\mu_i^* - r^* = \text{cov}(\frac{\delta_i}{S_i^*}, \mu_p^{'}(\delta - C_N^*)) \approx \lambda_\delta^* \text{cov}(\frac{\delta_i}{S_i^*}, \delta) - \lambda\varepsilon_N^* \text{cov}(\frac{\delta_i}{S_i^*}, \varepsilon_N),\ i=1,...,L$$

where

$$\lambda_\delta^* = -\mu^{''}(\mu_\delta - C_N^E)\{\frac{\mu_N^{''} C_N^E + y_N \mu_p^{''}(\mu_\delta - C_N^E)}{\mu_N^{''} C_N^E - y_N \mu_p^{''}(\mu_\delta - C_N^E)(C_N^E - \mu_{\varepsilon N}) + 2 y_N \mu_p^{''}(\mu_\delta - C_N^E)}\}$$

$$\lambda_{\varepsilon N}^* = \frac{y_N \mu_p^{''}(\mu_\delta - C_N^E)}{\mu_N^{''} C_N^E + y_N \mu_p^{''}(\mu_\delta - C_N^E)} \lambda_\delta^*$$

$r^* = 1/E[\mu_p^{'}(\delta - C_N^*)] - 1$, $\mu_\delta = E[\delta]$, $\mu_{\varepsilon N} = E[\varepsilon_N]$ and C_N^E is the solution to

$$\mu_N^{'}(C_N^E) = y_N[\mu_p^{'}(\mu_\delta - C_N^E) - \mu_p^{''}(\mu_\delta - C_N^E) \bullet (C_N^E - \mu_{\varepsilon N})].$$

Basak also extended the asset pricing into a dynamic model in which the non-price-taker's optimal consumption portfolio strategy is time-inconsistent in the sense that the non-price-taker has an incentive to deviate at a later date from the strategy chosen at time 0. The equilibrium is as follows:

$$\mu_i^*(t) - r^*(t) = \frac{\lambda_\delta^*(t)}{\mu_p^{'}(\delta(t) - C_N^*(t))} \text{cov}(\frac{dS_i^*(t)}{S_i^*(t)}, d\delta(t)) -$$

$$\frac{\lambda_{\varepsilon N}^*(t)}{\mu_p^{'}(\delta(t) - C_N^*(t))} \text{cov}(\frac{dS_i^*(t)}{S_i^*(t)}, d\varepsilon_N(t))$$

In addition to the aggregate consumption, the non-price-taking-taker's endowment is an extra factor driving the equilibrium allocations and prices.

8.3.7 Market frictions

Much of the theory and empirical studies on asset pricing were conducted in a framework of frictionless asset markets. Agents are assumed to be able to buy and sell a given security at the same price and without paying any transaction costs. However, this is not true in the real world asset markets.

Heaton and Lucas (1992) examined a three period asset pricing model. They assumed that individuals have access to a limited set of securities markets and face aggregate and individual uncertainty. They showed that asset prices vary predictably with the assumed market structure for trading frictions and this gives an implication that market incompleteness can explain asset market behavior. Later, Heaton and Lucas (1996) examined asset prices and consumption patterns in a model in which agents face both aggregate and idiosyncratic income shocks, and insurance markets are incomplete. Agents reduce consumption variability by trading in a stock and bond market to offset idiosyncratic shocks, but frictions in both markets limit the extent of trade. Transaction costs in the stock and bond markets generate an equity premium and lower the risk-free rate. They decomposed the effect of transaction costs on the equity premium into two components: a direct effect and an indirect effect. However, the model can not explain the observed second-moment differentials.

He and Modest (1995) considered market frictions of short sales, borrowing, solvency and trading cost. They performed diagnostic tests for consumption-based asset pricing models and showed that none of the market frictions alone-with the possible exception of solvency constraints can explain the rejection of the first order equilibrium conditions between the consumption and the asset returns. However, a combination of short sale and borrowing constraints and trading costs does not yield a rejection of the model.

Luttmer (1996) examined how proportional transaction costs, short-sale constraints, and marginal requirements affect inferences based on asset return data about intertemporal marginal rates of substitution (IMRSs). He showed that small transaction costs can greatly reduce the required variability of IMRSs. When the possibility of costless selling is eliminated, T-bills short provide a potential solution to the so-called "short rate puzzle".

8.4 Tests of CAPM

Early empirical investigations of the CAPM were based on the natural implications that arise from the theory: higher returns should be expected from stocks that have higher Betas, and the relationship between expected return and Beta should be linear. In the case of the Sharpe-Lintner-Mossin CAPM, the slope of this line should be equal to the market risk premium, and the intercept should be equal to the risk-free rate. For the zero-Beta CAPM, the slope should be less than the market risk premium, while the intercept should be greater than the risk-free rate. Moreover, there should be no systematic reward for bearing non-market risk, and any deviations in realized returns from the CAPM should not be predictable.

Black *et al.* (1972) and Fama and MacBeth (1973) used the two-pass methodology in which the first pass gives the estimates of portfolio Betas and the second pass regresses average returns on the estimated Betas from the first pass. However, Roll (1977) demonstrated that the market is not a single equity market, but an index of all the wealth. He pointed that the portfolio used by Black *et al.* is not the true portfolio. Roll showed that unless the market portfolio is known with certainty then the CAPM can never be tested.

Responses to the Roll's critique, Shanken (1987) and Kandel and Stambaugh (1987) argued that even though the stock market is not the true market portfolio, it must nevertheless be highly correlated with the true market. Another response is the use of proxies that include broader sets of assets such as bonds and property. However, Stambaugh (1982) found that even when bonds and real estate are included into the market proxy, the CAPM is still rejected.

Recent empirical tests of CAPM concentrate on other risk factors that affect stock returns such as book-to-market equity [Fama and Frenceh (1992)]. Some investigations have been done to see whether CAPM can be applied to un-mature financial markets (see Xia, Wang and Deng (1999)).

8.5 An Intertemporal Capital Asset Pricing Model

A key assumption in Markowitz portfolio optimization and the original CAPM is that agents make their decisions for only one time period. This is clearly an unrealistic assumption since investors can and do rebalance their portfolios on a regular basis.

Merton (1973b) developed an equilibrium model of the capital market which has the simplicity and empirical tractability of the capital asset pricing model and is consistent with the expected utility maximization and the limited liability of assets. The model also provides a specification of the relationship among yields that is more consistent with empirical evidence.

Merton assumed that the capital market was structured as follows:

1) All assets have limited liability.

2) There are no transaction costs, taxes, or problems with indivisibilities of assets.

3) There are a sufficient number of investors with comparable wealth levels so that each investor believes that he can buy and sell as much of an asset as he/she wants at the market price.

4) The capital market is always in equilibrium (*i.e.*, there is no trading at non-equilibrium prices).

5) There exists an exchange market for borrowing and lending at the same rate of interest.

6) Short sales of all assets, with full use of the proceeds, are allowed.

7) Trading in assets takes place continually in time.

8) The vector set of stochastic processes describing the opportunity set and its changes, is a time-homogeneous Markov process.

9) Only local changes in the state variables of the process are allowed.

10) For each asset in the opportunity set at each point in time t, the expected rate of return per unit time, defined by

$$\alpha = E_t[(P(t+h) - P(t))/P(t)]/h$$

and the variance of the return per unit time, defined by

$$\sigma^2 = E_t[([P(t+h) - P(t)]/P(t) - \alpha h)^2]/h,$$

are finite with $\sigma^2 > 0$, and are (right) continuous functions of h, where " E_t " is the conditional expectation operator, conditional on the levels of the state variables at time t, as h tends to zero, α is called the instantaneous expected return and σ^2 the instantaneous variance of the return.

With the above assumptions, the equilibrium relationship between the expected return on an individual asset and the expected return on the market can be expressed as follows:

$$\alpha_i - r = \frac{\sigma_i[\rho_{iM} - \rho_{in}\rho_{nM}]}{\sigma_M(1-\rho_{nM}^2)}(\alpha_M - r) + \frac{\sigma_i[\rho_{in} - \rho_{iM}\rho_{nM}]}{\sigma_n(1-\rho_{Mn}^2)}(\alpha_n - r),\ i=1,\dots,n\text{-}1$$

where α_i and α_M are respectively the expected return of asset i and the expected return of market index M. σ_i and σ_M are standard deviations of return of asset i and market index M. r is the riskless lending and borrowing rate. ρ is the correlation coefficient.

The above equation states that, in equilibrium, investors are compensated in terms of expected return, for market (systematic) risk and for bearing the risk of unfavorable (from the point of view of the aggregate) shifts in the investment opportunity set. The model is a natural generalization of the security market line of the classical capital asset pricing model. Note that if a security has no market risk (*i.e.*, $\beta_i = 0 = \rho_{iM}$), its expected return will not be equal to the riskless rate as forecasted by the usual model.

However, the assumptions, principally homogeneous expectations, which it holds in common with the classical model, make the new model subject to some criticisms. Moreover, the above intertemporal CAPM with stochastic investment opportunities states that the expected excess return on any asset is given by a 'multi-Beta' version of the CAPM with the number of Betas being equal to one plus the number of state variables needed to describe the relevant characteristics of the investment opportunity set.

There are some other progress both in theory and empirical tests about the intertemporal asset pricing model. Campbell (1993) used a loglinear approximation to the budget constraint to substitute out consumption from a standard intertemporal asset pricing model. Zhou (1998) presented a multi-asset intertemporal general equilibrium model of portfolio selection and asset pricing with differential information. Malliaris and Stein (1999) presented a general intertemporal price determination model. The model incorporates the "random walk" hypothesis which assumes that price volatility is exogenous and unexplained. The main difference of the model is that it considers the learning process when more information is available. Magill and Quinzii (2000) derived an equilibrium of an infinite-horizon discrete-time CAPM economy in which agents have discounted expected quadratic utility functions. They showed that there is an income stream obtainable by trading on the financial markets which best approximates perfect consumption smoothing (called the least variable income stream or LVI) such that the equilibrium consumption of each agent is some multiple of the LVI and some share of aggregate output.

Chen (1986) presented an intertemporal capital asset pricing model with heterogeneous beliefs. The paper shows that an asset's risk consists of three components: the market consensus of volatility risk, the market consensus of the risk induced by changes in the investment opportunity set, and risk associated with uncertain shifts in the investors' subjective expectations. The equilibrium model derived in the paper under heterogeneous beliefs provides a generation of the CAPM and Merton's intertemporal capital asset pricing model.

8.6 Consumption-Based Capital Asset Pricing Model

Breeden (1979) pointed that Merton's intertemporal capital asset pricing model is quite important from a theoretical standpoint and that it is not very tractable for empirical testing, nor is very useful for financial decision-making. Breeden showed that the Merton's multi-Beta pricing equation can be collapsed into a single-beta equation, where the instantaneous expected excess return on any security is proportional to its 'Beta' with respect to aggregate consumption alone.

Denote C_t as the growth rate in aggregate consumption per capital at time t and R_{it} as the rate of return on asset i in period t. We have

$$R_{it} = \alpha_t + \beta_i C_t + e_{it}$$

where $E(e_{it}) = 0$, and the covariance between residuals and the index is zero, *i.e.*,

$$\operatorname{cov}(e_{it}, C_t) = 0$$

Then, the equilibrium condition is

$$E(R_i) = E(R_Z) + \gamma_1 \beta_1$$

where γ_1 is the market price of the consumption Beta and $E(R_Z)$ is the expected return on a portfolio with zero consumption Beta.

Further work about consumption-based CAPM refers to Lewis (1991), and Jagannathan (1985). Empirical tests about the consumption-based CAPM can be found in Ghyseles and Hall (1990) , Hansen and Singletion (1982) and Dunn and Singleton (1986).

8.7 Arbitrage Pricing Model

Unlike the Sharpe's single index model, Ross (1976) proposed the arbitrage pricing theory in which the model assumes that there are multiple factors to represent the fundamental risks in the economy. The multiple factors allow an asset to have serveral measures of the systematic risk. Each measure captures the sensitivity of the asset to the corresponding pervasive risk.

Ross made the following assumptions:

1) Liability limitations: there exists at least one asset with limited liability in the sense that there is some bound to the losses for which an agent is liable.

2) Non-negligibility of type B agents: there is at least one type B agent which is not asymptotically negligible. type B agent refers to the agents whose coefficients of relative risk aversion are uniformly bounded, *i.e.,* $\sup_x\{-(U''(x)x/U'(x)\}<\infty$. An agent is called negligible if $w^v/w\to 0$, where w^v is the agent's wealth and w is the total wealth.

3) Homogeneity of expectations: all agents hold the same expectations E (expected return vector) and all agents are risk-averse.

4) Extent of disequilibria: Let ξ^i be the aggregate demand for the i-th asset as a fraction of total wealth. $\xi^i\geq 0$.

5) Bounded expectation: the expectation of E_i is bounded.

Under these assumptions, Ross showed that the risk premium on an asset is the β weighted sum of the risk premiums factors:

$$E_i=\rho+\beta_{i1}(E^1-\rho)+\cdots+\beta_{ik}(E^k-\rho),$$

where E^l is the return on all portfolios with $\alpha\beta^s=0$ for all $s\neq l$ and $\alpha\beta^l=1$. α is the portfolio of risky assets (α_i is the proportion of wealth placed in the ith risky asset). $\beta^s=[\beta_{1s},\cdots,\beta_{ns}]'$. ρ is the return on zero-Beta portfolios.

Huberman (1982) gave the definition of arbitrage and presented a simple approach to the arbitrage pricing theory by the argument of no-arbitrage. The arbitrage pricing theory implies that if asset returns have a factor structure, then an approximate multi-Beta representation holds with respect to the factors as reference variables.

Reisman (1992a) assumed that asset returns satisfy a factor structure and derived a condition that the approximate multi-Beta representation holds with respect to a set of reference variables, but these reference variables may not be the factors. The paper shows that as long as there exists an approximate factor structure for returns, almost any set of variables correlated with the factors can serve as the benchmark in an approximate APT expected return relation. Later, Shanken (1992) gave a simple proof of the result.

Some empirical tests about APT can be found in Roll and Ross (1980), Shanken (1982, 1985) and Dybvig and Ross (1982).

8.8 Intertemporal Arbitrage Pricing Theory

Just like that the CAPM has been extended to the multi-period case, some efforts were made to generalize the single period arbitrage pricing theory.

Ohlson and Garman (1980) put the security prices as endongenous functions of states and made the following assumptions:

1) Investors have homogenous Markovian beliefs, F, and agree upon the prices and dividends that will obtain for any given state z_t (state variable).

2) Equilibrium asset prices are determined by

$$a_i(z_t) - RP_i(z_t) = \sum_{k=1}^{K} \lambda_k b_{ik}(z_t)$$

where R is the riskfree asset and $\lambda_1, \lambda_2 \cdots \lambda_K$ are the constant for the Ross's single period APT. R and $\lambda_1, \lambda_2 \cdots \lambda_K$ are exogenous and state-independent.

$$a_i(z_t) = E[P_i(Z_{t+1}) + D_{i,t+1} \mid z_t] = P_i(z_t)\alpha_i(z_t)$$

$$b_{ik}(z_t) = \mathrm{cov}[P_i(Z_{t+1}) + D_{i,t+1}, \delta_{k,t+1} \mid z_t] = P_i(z_t)\beta_{ik}(z_t)$$

where P and D denote price and dividends respectively, $\delta_{k,t+1}$ is random variable associated with the single period model.

3) The state at time t is given by $z_t = (d_{1,t}, d_{2,t}, \cdots)$ and the state dynamics is specified by

$$D_{i,t+1} = (1 + \theta_i + \sum_{k=1}^{K} w_{ik}\delta_{k,t+1} + \varepsilon_{i,t+1})d_{i,t}$$

where D and d are dividends in which d is known and D is random variable. $\delta_{1,t},\cdots,\delta_{K,t},\varepsilon_{1,t},\varepsilon_{2,t},\cdots$ are cross-sectionally and intertemporally independent with zero means and finite variance.

With the above assumptions, Ohlson and Graman showed that the return is as follows:

$$\alpha_i(z_t) = \alpha_i = (1 + \theta_i)\Gamma_i$$

$$\Gamma_i = R(1 + \theta_i - \sum_{k=1}^{K} \lambda_k w_{ik})^{-1}$$

Reisman (1992b) generalized the one-period APT and considered a continuous trading economy to show that the arbitrage pricing theory holds in each infinitesimal period.

8.9 Comparison of CAPM and APT

Between CAPM and APT, which model is more helpful for financial decision making? Actually, both of these two models have its advantages and disadvantages.

The APT implies the same expected return-Beta relationship as the CAPM yet does not require that all investors be mean-variance optimizers. The price of this generality is that the APT does not guarantee this relationship for all securities at all times.

Sharpe (1998) pointed out that " The APT is stronger in that it makes some very strong assumptions about the return-generating process, and it is weaker because it does not tell you very much about the expected return on those factors. The CAPM and its extended versions offer some notion of how people with preferences determine prices in the market. The CAPM tells you more. The CAPM does not require that there are three factors or five factors. There could be a million. Whatever number of factors there may be, the expected return of a security will be related to its exposure to those factors."

8.10 Conclusions

In this chapter, we have illustrated the standard capital asset pricing theory and its serveral kinds of derivatives. For multi-period asset pricing, an intertemporal capital asset pricing model and some of its extensions are discussed. Also the arbitrage pricing model and its intertemporal form are described.

Although there are a lot of publications on the asset pricing theory and the single period CAPM is indeed widely used in the financial markets to calculate the price of assets. However, some further investigations still need to be done:

(1) Relaxing the assumptions of the CAPM: what's the risk premium in markets with transactions costs as well as other factors such as one big investor could affect the price of some securities.

(2) Multi-period asset pricing: most investors are based on a multi-period bases to maximize their utilities in the long run. Currently, the intertemporal CAPM are not very helpful for decision making. Thus there is still some work needed to be done for the intertemporal CAPM,

(3) How to determine the risk premium for assets in un-mature financial markets? Un-mature financial markets usually don't satisfy the linear relationship of the CAPM. Then determining the right relationship is very useful for those markets.

9. Empirical Tests of CAPM for China's Stock Markets

9.1 Introduction

Originally proposed by Sharpe (1964), Lintner (1965) and Black (1972) , the Capital Asset Pricing Model has long shaped the way academics and practitioners think about average returns and risk. The essential prediction of the model is that the market portfolio of invested wealth is mean-variance efficient in the sense of Markowitz (1959).

The CAPM is an equilibrium model under the assumption of a perfect market. Under this assumption, any investor in the stock market owns the market portfolio which is a portfolio in which the fraction invested in any asset is equal to the market value of that asset divided by the market value of all the risky assets. Each investor adjusts the risk of the market portfolio with lending or borrowing at the riskless rate.

There are many empirical results of the CAPM which support the efficiency of the CAPM , the efficiency of the market portfolio and the relationship between stocks' expected returns and their market Betas: (1) expected returns on securities are positive linear functions of their market Betas (the slope in the regression of a security's return on the market's return) and (2) market Betas suffice to describe the cross-section of expected returns. Fama and Macbeth (1973) tested the two-parameter portfolio model and models of market equilibrium derived from the two-parameter portfolio model. Their results show that the pricing of common stocks reflects the attempts of risk-averse investors to hold portfolios that are efficient in terms of expected value and dispersion of risk. Black *et al.* (1972) tested the asset pricing theory using the technique of group portfolio and found that there was a positive simple relation between average stock returns and Beta. Blume *et al.* (1973) also reported evidence consistent with the mean-variance efficiency of the market portfolio.

However, there are also some empirical contradictions of validity of CAPM. Banz (1981) observed that market equity, ME (a stock's price times share's outstanding), added to the explanation of the average returns in addition to the market Beta. Average returns on small (low ME) stocks are too high given their Beta estimates, and average returns on large stocks are too low. Bhandari (1988) mentioned that there exist positive relations between leverage and average return, and thus he concluded that leverage also helps explaining the average stock returns. Stattman (1980) and Rosenberg *et al.* (1985) found that the ratio of a firm's book value of common equity to its market value can explain the average returns on U.S. stocks. Chan *et al.* (1991) considered the book-to-market equity and found that the average returns on the Japanese stocks have a strong relationship with the book-to-market value. Basu (1983) tested the U.S. stocks to consider earnings-price ratios, size and market Beta and showed that earnings-price can also help to explain the average returns of those stocks' average returns.

All those empirical results urge us to see whether the results derived from the U.S. stock market and the Japanese stock market are similar to or different from those derived from the stock markets of China. Another motivation for this study is that there are too few results about non-standard forms of the CAPM which consider the taxes and dividend effects.

The models to be tested are given in Section 9.2. The data and methodology are illustrated in Section 9.3. Empirical results and some discussions about the results are presented in Section 9.4. Finally, a few conclusion remarks are given in Section 9.5.

9.2 Models

If in a frictionless market, investors have homogeneous expectations and can unlimitedly lend and borrow money at the riskless rate, then the Sharp-Lintner version model of CAPM can be formulated as follows:

$$E(R_i) = R_f + \beta_{im}(E(R_m) - R_f)$$

$$\beta_{im} = \frac{Cov(R_i, R_m)}{Var(R_m)}$$

where R_m is the return on the market portfolio and R_f is the return rate of riskfree asset.

There exist many empirical tests of this version model for different stock markets [Sharpe and Cooper (1992)]. However, in this chapter, similar to Black *et al.* (1972) and Fama and MacBeth (1973), we consider a two-parameter version model of CAPM to test the stock markets of China since the assumption that investors can lend and borrow money at the riskfree rate with unlimited quantity is untenable in China.

9.2.1 Two parameter model of CAPM

In the absence of a riskfree asset, Black (1972) derived the following more general version of CAPM:

$$E(R_i) = E(R_0) + \beta_{im}(E(R_m) - E(R_0))$$

where R_i is the return of asset i, R_m is the return on the market portfolio and R_0 is the zero-Beta portfolio.

The model describes that the expected return of asset i in excess of the zero-Beta stock return is linearly related to its Beta. In order to test this model, we formulate the testable stochastic model as follows:

$$E(R_i)=\hat{\gamma}_0 + \hat{\gamma}_1 \hat{\beta}_i + \hat{\gamma}_2 \hat{\beta}_i^2 + \hat{\gamma}_3 \hat{S}(\varepsilon_i)$$

where $S(\varepsilon_i)$ is unsystematic risk which will not contribute to the risk premium of asset i if the two parameter model holds.

The testable implications of this model are:

(1) Linearity: if the linear relationship of excess return of asset i with its Beta is true, then $E(\hat{\gamma}_2)$ =0;

(2) No-systematic effects of non-beta risk: if the risk premium is only due to the market Beta, then $E(\hat{\gamma}_3)$=0;

(3) Positive expected return-risk tradeoff: if the excess return of asset i is positive to its market Beta, then $E(\hat{\gamma}_1) > 0$.

9.2.2 CAPM with taxes on dividends

In the absence of taxes, capital asset pricing theory suggests that individuals choose mean-variance efficient portfolios. With the taxes on dividends, individuals would be expected to choose portfolios that are mean-variance efficient in after-tax rates of return.

Brennan (1973) first proposed an extended form of the single period capital asset pricing model that accounts for the taxation of dividends. Under the assumption of proportional individual tax rates (not a function of income), certain dividends, and unlimited borrowing at the riskless rate of interest, he derived the following equilibrium relationship:

$$E(R_i) - r_f = b\beta_i + \tau(d_i - r_f)$$

where R_i is the security i 's before tax total return, β_i is its systematic risk, $b = [E(R_m - r_f) - \tau(d_i - r_f)]$ is the after tax excess rate of return on the market portfolio, r_f is the return on a riskless asset, d_i is the dividend yield on security i, the subscript m denotes the market portfolio, and τ is a positive coefficient that accounts for the taxation of dividends and interest as ordinary income and taxation of capital gains at a preferential rate. However, with this model, when interest on borrowing exceeds dividend income, the investor would pay a negative tax due to the assumption of unlimited borrowing and of constant tax rates which may vary across individuals.

In this chapter, we also consider the model proposed in Litzenberger and Ramaswamy (1979) with constraints on borrowing and on a progressive tax scheme. We test the following stochastic model:

$$E(R_p) = \hat{\gamma}_0 + \hat{\gamma}_1 \hat{\beta}_i + \hat{\gamma}_2 \hat{\beta}_i^2 + \hat{\gamma}_3 \hat{S}(\varepsilon_i) + \hat{\gamma}_4 d_i$$

If there is no effects of dividend taxes, $E(\hat{\gamma}_4)$ should be equal to zero.

9.3 Data and Methodology

The data for this study are monthly percentage returns for all the stocks of type A and type B traded on the Shanghai Stock Exchange and the Shenzhen Stock Exchange during the period October 1996 through December 1998. We use the corresponding stock index to represent the market portfolio. One reason to choose October 1996 as the beginning is that most of the data of type B stocks are available from this period. The number of stocks in our empirical tests is presented in Table 9.1.

Table 9.1 Number of stocks in the empirical test

	SHA	SZA	SHB	SZB
Number of stocks	276	185	42	42

Note: SHA: Stock type A listed in the Shanghai Stock Exchange;

SZA: Stock type A listed in the Shenzhen Stock Exchange;

SHB: Stock type B listed in the Shanghai Stock Exchange;

SZB: Stock type B listed in the Shenzhen Stock Exchange.

In our empirical test, the ordinary least square (OLS) method is applied to estimate the parameters. t test and F test are used to see whether these parameters are significant.

In order to reduce the dependency of regression residual errors, we use group techniques to form portfolios and thus this will reduce the asymptotic bias during the estimates. Following are the six steps in our implementation:

(1) Implement the first pass regression to calculate the estimated value β_i as well as the standard error of regression S_i of each stock i with the following equation:

$$R_{it} = a_i + \beta_i R_{mt} + \varepsilon_i , \; t = 1,\cdots,T$$

(2) Rank β_i in ascending order and divide them into twenty groups, calculate the average β of each group:

$$\beta_p = \frac{1}{n_p} \sum_{j=1}^{n_p} \beta_j$$

where n_p is the stock number of p-th portfolio, $p = 1,\cdots,20$.

(3) For each group of stocks, calculate average return as follows:

$$R_p = \frac{1}{n_p} \sum_{j=1}^{n_p} R_j$$

(4) Calculate average standard regression error of each group:

$$S_p = \frac{1}{n_p} \sum_{j=1}^{n_p} S_j$$

(5) Regress different models in different stock markets.

(6) Apply t test and F test to examine whether the regression parameters are significant.

9.4 Empirical Results and Discussions

During the first pass regression, we formed twenty portfolios for both two stock markets of China and for both type A stocks and type B stocks as the following tables:

Table 9.2 20 portfolios of type A stocks in Shanghai Stock Exchange

P_i	$E(R_i)$	β_i	$S(\varepsilon_i)$	P_i	$E(R_i)$	β_i	$S(\varepsilon_i)$
1	0.014	0.203	0.138	11	0.003	1.034	0.105
2	0.017	0.527	0.137	12	0.017	1.081	0.123
2	0.014	0.641	0.118	13	0.012	1.125	0.109
4	- 0.007	0.699	0.122	14	0.012	1.174	0.109
5	0.003	0.756	0.108	15	0.020	1.263	0.122
6	- 0.003	0.806	0.118	16	0.011	1.350	0.135
7	0.013	0.847	0.117	17	0.015	1.445	0.126
8	0.007	0.887	0.122	18	0.020	1.570	0.124
9	0.005	0.945	0.095	19	0.019	1.724	0.138
10	0.004	0.991	0.117	20	0.035	2.052	0.153

Note: $E(R_i)$: the average return of portfolio i ; $S(\varepsilon_i)$: the average residual risk of portfolio i .

Table 9.3 20 portfolios of type A stocks in Shenzhen Stock Exchange

P_i	$E(R_i)$	β_i	$S(\varepsilon_i)$	P_i	$E(R_i)$	β_i	$S(\varepsilon_i)$
1	- 0.007	0.365	0.128	11	0.022	1.034	0.139
2	0.006	0.534	0.128	12	0.018	1.074	0.142
2	- 0.003	0.641	0.122	13	0.013	1.122	0.114
4	- 0.003	0.690	0.112	14	0.009	1.179	0.098
5	- 0.003	0.763	0.124	15	0.032	1.277	0.141
6	0.012	0.814	0.148	16	0.028	1.306	0.130
7	0.002	0.869	0.100	17	0.031	1.389	0.125
8	0.006	0.920	0.125	18	0.038	1.441	0.151
9	0.015	0.960	0.129	19	0.029	1.542	0.143
10	0.006	0.998	0.113	20	0.038	1.755	0.157

Table 9.4 20 portfolios of type B stocks in Shanghai Stock Exchange

P_i	$E(R_i)$	β_i	$S(\varepsilon_i)$	P_i	$E(R_i)$	β_i	$S(\varepsilon_i)$
1	- 0.001	0.321	0.170	11	- 0.014	0.650	0.116
2	- 0.012	0.420	0.158	12	- 0.037	0.708	0.112
2	0.006	0.447	0.165	13	- 0.023	0.736	0.125
4	- 0.006	0.475	0.157	14	- 0.017	0.771	0.144
5	- 0.005	0.521	0.166	15	- 0.025	0.810	0.114
6	0.003	0.545	0.170	16	- 0.027	0.840	0.142
7	- 0.005	0.553	0.151	17	- 0.003	0.891	0.145
8	- 0.013	0.569	0.126	18	- 0.001	0.933	0.189
9	- 0.028	0.621	0.155	19	- 0.003	1.084	0.167
10	- 0.026	0.634	0.160	20	0.003	1.134	0.144

Table 9.5 20 portfolios of type B stocks in Shenzhen Stock Exchange

P_i	$E(R_i)$	β_i	$S(\varepsilon_i)$	P_i	$E(R_i)$	β_i	$S(\varepsilon_i)$
1	0.091	- 0.016	0.667	11	- 0.028	0.683	0.178
2	- 0.020	0.540	0.117	12	- 0.017	0.699	0.186
2	- 0.032	0.556	0.098	13	- 0.007	0.707	0.132
4	- 0.022	0.565	0.180	14	0.000	0.725	0.148
5	- 0.032	0.576	0.180	15	- 0.017	0.752	0.150
6	- 0.019	0.583	0.132	16	- 0.022	0.793	0.121
7	- 0.016	0.609	0.133	17	- 0.019	0.822	0.168
8	- 0.020	0.632	0.127	18	- 0.032	0.830	0.196
9	0.008	0.650	0.183	19	- 0.003	0.913	0.190
10	- 0.015	0.668	0.153	20	0.011	1.183	0.273

From the above tables, one can observe that it is not the case that large Betas will have large expected returns for the different stock types listed in the two different Exchanges.

We implement the second pass regression to derive the results as follows:

Table 9.6 Regressing equation $R_p = \hat{\gamma}_0 + \hat{\gamma}_1 \hat{\beta}_i + \hat{\gamma}_2 \hat{\beta}_i^2 + \hat{\gamma}_3 \hat{S}(\varepsilon_i) + \hat{\eta}_i$

	SHA	SZA	SHB	SZB
$\hat{\gamma}_0$	- 0.006	- 0.068	0.007	0.012
$\hat{\gamma}_1$	- 0.016	0.048	- 0.162	- 0.131
$\hat{\gamma}_2$	0.012	- 0.009	0.111	0.090
$\hat{\gamma}_3$	0.157	0.335	0.239	0.112
$S(\hat{\gamma}_0)$	0.030	0.014	0.036	0.479
$S(\hat{\gamma}_1)$	0.020	0.017	0.069	0.114
$S(\hat{\gamma}_2)$	0.020	0.010	0.046	0.077
$S(\hat{\gamma}_3)$	0.186	0.085	0.117	0.084
$t(\hat{\gamma}_0)$	0.203	4.836	0.183	0.229
$t(\hat{\gamma}_1)$	0.795	2.789	2.334	1.157
$t(\hat{\gamma}_2)$	1.190	0.850	2.416	1.161
$t(\hat{\gamma}_3)$	0.840	3.944	2.405	1.353
F	9.324	41.142	9.416	23.032

Table 9.7 Regressing equation $R_p = \hat{\gamma}_0 + \hat{\gamma}_1 \hat{\beta}_i + \hat{\gamma}_2 \hat{\beta}_i^2 + \hat{\eta}_i$

	SHA	SZA	SHB	SZB
$\hat{\gamma}_0$	0.018	- 0.019	0.066	0.082
$\hat{\gamma}_1$	- 0.028	0.026	- 0.230	0.279
$\hat{\gamma}_2$	0.018	0.005	0.155	0.189
$S(\hat{\gamma}_0)$	0.245	0.363	0.022	0.010
$S(\hat{\gamma}_1)$	0.445	0.716	0.066	0.033
$S(\hat{\gamma}_2)$	0.189	0.333	0.045	0.026
$t(\hat{\gamma}_0)$	0.245	0.051	2.961	8.250
$t(\hat{\gamma}_1)$	0.063	0.036	3.473	8.403
$t(\hat{\gamma}_2)$	0.095	0.015	3.475	7.159
F	12.665	28.756	9.160	29.181

Table 9.8 Regressing equation $R_p = \hat{\gamma}_0 + \hat{\gamma}_1 \hat{\beta}_i + \hat{\gamma}_3 S(\varepsilon_i) + \hat{\eta}_i$

	SHA	SZA	SHB	SZB
$\hat{\gamma}_0$	- 0.039	- 0.056	- 0.068	- 0.047
$\hat{\gamma}_1$	0.009	0.031	0.002	- 0.001
$\hat{\gamma}_3$	0.338	0.299	0.371	0.203
$S(\hat{\gamma}_0)$	0.014	0.010	0.020	0.443

$S(\hat{\gamma}_1)$	0.005	0.010	0.010	0.479
$S(\hat{\gamma}_3)$	0.122	0.077	0.117	0.887
$t(\hat{\gamma}_0)$	2.786	5.610	3.420	0.107
$t(\hat{\gamma}_1)$	9.300	6.200	0.220	0.003
$t(\hat{\gamma}_3)$	2.769	3.860	3.170	0.229
F	11.986	54.747	8.707	30.069

Table 9.9 Regressing equation $R_p = \hat{\gamma}_0 + \hat{\gamma}_1 \hat{\beta}_i + \hat{\eta}_i$

	SHA	SZA	SHB	SZB
$\hat{\gamma}_0$	- 0.002	- 0.024	- 0.010	0.030
$\hat{\gamma}_1$	0.013	0.037	- 0.003	- 0.060
$S(\hat{\gamma}_0)$	0.486	0.483	0.314	0.574
$S(\hat{\gamma}_1)$	0.427	0.443	0.438	0.812
$t(\hat{\gamma}_0)$	0.004	0.049	0.031	0.052
$t(\hat{\gamma}_1)$	0.030	0.083	0.007	0.074
F	12.914	28.334	8.523	11.226

If we have the probability 95% as the confidence level, only the following equations can pass both the t test and F test.

$$\text{SHA:}\ R_i = -0.039 + 0.009\,\hat{\beta}_i + 0.338\,S(\varepsilon_i) + \hat{\eta}_i\,,$$

$$\text{SZA: } R_i = -0.056 + 0.031\,\hat{\beta}_i + 0.299\,S(\varepsilon_i) + \hat{\eta}_i\,,$$

$$\text{SHB: } R_i = 0.066 - 0.230\,\hat{\beta}_i + 0.155\,\hat{\beta}_i^2 + \hat{\eta}_i$$

and

$$\text{SZB: } R_i = 0.082 + 0.279\,\hat{\beta}_i + 0.189\,\hat{\beta}_i^2 + \hat{\eta}_i$$

From the above equations, we observe that there does not exist linear relationships between the expected returns of stocks and the market Betas as reported in some developed markets such as the New York Stock Exchange. For stock type A in both the Shanghai Stock Exchange and the Shenzhen Stock Exchange, the risk premium is linear with the market Beta as well as the unsystematic risk which can not be eliminated through diversification. For stock type B listed in both the two Stock Exchanges, the expected returns of stocks have a non-linear relationship with the market Betas.

Actually, we are not surprised with the above empirical result that the capital asset pricing model does not hold for the stock markets of China. CAPM is an equilibrium model which assumes that there are no friction in the stock markets and all investors have the homogeneous expectations. Regardless of the homogeneous expectations of investors, the assumption of markets' non-friction does not hold in China. Compared with some mature markets such as the New York Stock Exchange which has a very long history, China's stock markets are still in their early stage with only a history of no more than ten years. The un-mature natures in China's stock markets contradict the assumptions of perfection which the equilibrium of CAPM holds.

To see the impact of dividend tax effects on stock returns, we consider the dividends as a parameter for regression. Because there are almost no dividends of type B stocks in both the Shanghai Stock Exchange and the Shenzhen Stock Exchange, we only consider the effects of dividend taxes on stock A type for both the two Stock Exchanges. We have the following results of regression:

Table 9.10 Regressing equation $R_p = \hat{\gamma}_0 + \hat{\gamma}_1\,\hat{\beta}_i + \hat{\gamma}_3\,S(\varepsilon_i) + \hat{\gamma}_4\,d_i + \hat{\eta}_i$

	$\hat{\gamma}_0$	$\hat{\gamma}_1$	$\hat{\gamma}_3$	$\hat{\gamma}_4$	$S(\hat{\gamma}_0)$	$S(\hat{\gamma}_1)$
SHA	- 0.040	0.009	0.342	0.568	0.014	0.010
SZA	- 0.054	0.031	0.289	- 0.850	0.010	0.010

continued

$S(\hat{\gamma}_3)$	$S(\hat{\gamma}_4)$	$t(\hat{\gamma}_0)$	$t(\hat{\gamma}_1)$	$t(\hat{\gamma}_3)$	$t(\hat{\gamma}_4)$	F
0.126	2.603	2.828	0.900	2.704	0.218	8.454
0.071	1.089	5.400	3.100	4.070	0.716	45.891

It is obviously that the parameter cannot pass the t test and we thus remove the dividend effect to derive the following regression results:

Table 9.11 Regressing equation $R_p = \hat{\gamma}_0 + \hat{\gamma}_1 \hat{\beta}_i + \hat{\gamma}_3 S(\varepsilon_i) + \hat{\eta}_i$

	$\hat{\gamma}_0$	$\hat{\gamma}_1$	$\hat{\gamma}_3$	$S(\hat{\gamma}_0)$	$S(\hat{\gamma}_1)$
SHA	- 0.040	0.009	0.341	0.014	0.007
SZA	- 0.056	0.031	0.299	0.010	0.007

continued

$S(\hat{\gamma}_3)$	$t(\hat{\gamma}_0)$	$t(\hat{\gamma}_1)$	$t(\hat{\gamma}_3)$	F
0.124	2.798	9.300	2.774	11.454
0.077	3.422	6.200	3.866	54.791

The above regressing model can pass both the t test and the F test, We have the following relationship of expected returns of stocks and the market Betas as well as the residuals:

$$\text{SHA: } R_i = -0.040 + 0.009\,\hat{\beta}_i + 0.341\,S(\varepsilon_i) + \hat{\eta}_i$$

and

$$\text{SZA: } R_i = -0.056 + 0.031\,\hat{\beta}_i + 0.299\,S(\varepsilon_i) + \hat{\eta}_i$$

The above regression coefficients are almost the same as those not considering the dividends. We can conclude that there are no effects of dividend taxes on the expected returns of stocks in the stock markets of China. This is not the same as the empirical results of some developed markets, where there is a great effect of dividend taxes to the expected returns of stocks. One obvious reason for the difference is that the dividends of stocks listed in China's markets are too small to be considered.

9.5 Conclusions

We have tested the two-parameter model of CAPM with and without the dividend tax effects. We discovered that the stock markets of China do not follow the CAPM as other very developed stock markets such as the New York stock market. Our results are full of interest. The market Betas of type A stocks listed in both China's two Stock Exchanges can not fully explain the expected returns of stocks. The risk premium of type A stocks in both the two stock markets of China are due not only to the market Betas, but also to the unsystematic risk of stocks. For the type B stocks listed in both the two Exchanges, there exists non-linear relationship between the expected returns of stocks and the market Betas.

However, one problem in our empirical tests of CAPM for the stock markets of China which we can not deny is that the sample of our tests is still not large enough although we have included all the stocks listed in both the two Stock Exchanges of China. This is largely due to that China's stock markets are at their early stages.

References

Akian, M., Menaldi, J.L. and Sulem, A. (1995), Multi-asset portfolio selection problem with transaction costs, *Mathematics and Computers in Simulation*, 38, 163-172.

Akian, M., Menaldi, J.L. and Sulem, A. (1996), On an investment-consumption model with transaction costs, *SIAM Journal on Optimization*, 34, 329-364.

Arditti, F.D. (1967), Risk and required return on equity, *Journal of Finance*, 22;19-36.

Arditti, F.D. (1971), Another look at mutual fund performance, *Journal of Financial and Quantitative Analysis*, 6, 909-912.

Arnott, R.D. and Wagner, W.H. (1990), The measurement and control of trading costs, *Financial Analysts Journal*, Nov/Dec, 73-80.

Atkinson, C. and Al-Ali, B. (1997), On an investment-consumption model with transaction cost: an asymptotic analysis, *Applied Mathematical Finance*, 4, 109-113

Ball, R. (1978), Anomalies in relationships between securities' yields and yield-surrogates, *Journal of Financial Economics*, 6, 103-126.

Ballestero, E. and Romero, C. (1996), Portfolio selection: a compromise programming solution, *Journal of Operational Research Society*, 47, 1377-1386.

Banz, R.W. (1981), The relationship between return and market value of common stocks, *Journal of Financial Economics*, 9, 3-18.

Basak, S. (1997), Consumption choice and asset pricing with a non-price-taking agent, *Economic Theory*, 10, 437-462.

Basak, S. (2000), A model of dynamic equilibrium asset pricing with heterogeneous beliefs and extraneous risk, *Journal of Economic Dynamics & Control*, 24, 63-95.

Basu, S. (1983), The relationship between earnings yield, market value, and return for NYSE common stocks: further evidence, *Journal of Financial Economics*, 12, 129-156.

Baumol, E.J. (1963), An expected gain-confidence limit criterion for portfolio selection, *Management Science*, 9, 174-182.

Bawa, V.S. (1975), Optimal rules for ordering uncertain prospects, *Journal of*

Financial Economics, 10, 849-857.

Bazaraa, M.S., Sherali, H.D. and Shetty, C.M. (1993), *Nonlinear Programming: Theory and Algorithms*, Second Edition, John Wiley & Sons, New York.

Bell, R. and Cover, T. (1980), Competitive optimality of logarithmic investment, *Mathematics of Operations Research*, 5, 161-166.

Bell, R. and Cover, T.M. (1988), Game-theoretic optimal portfolios, *Management Science*, 34, 724-733.

Bernardo, A.E. and Ledoit, O. (2000), Gain, loss and asset pricing, *Journal of Political Economics*, 108, 144-172.

Bhandari, L.C. (1988), Debt/equity ratio and expected common stock returns: empirical evidence, *Journal of Finance*, 43, 507-528.

Bicchieri, C. (1993), *Rationality and Coordination*, Cambridge University Press, London.

Bicksler, J.L. (1979), *Handbook of Financial Economics*, North-Holland, Amsterdam.

Black, F. (1972), Capital market equilibrium with restricted borrowing, *Journal of Business*, 45, 444-455.

Black, F., Jensen, M. and Scholes, M. (1972), The capital asset pricing model: some empirical results, in *Studies in the Theory of Capital Markets* (ed. by M. Jensen), Praeger, New York.

Black, F. and Scholes, M. (1973), The pricing of options and corporate liabilities, *Journal of Political Economics*, 81, 637-659.

Blume, M. (1975), Betas and their regression tendencies, *Journal of Finance*, 10, 785-795.

Blume, M. and Friend, I. (1973), A new look at the capital asset pricing model, *Journal of Finance*, 28, 19-33.

Blum, A. and Kalai, A. (1997), Universal portfolios with and without transaction costs, in *Proceedings of the Tenth annual Conference on Computational Learning Theory,* ACM Press, New York, 309-313.

Breeden, D. (1979), An intertemporal asset pricing model with stochastic consumption and investment opportunities, *Journal of Financial Economics*, 7, 265-296.

Brennan, M.J. (1973), Taxes, market valuation and corporate financial policy, *National Tax Journal*, 23, 4178-427.

Brennan, M.J. (1975), The optimal number of securities in a risky asset portfolio when there are fixed costs of transaction: theory and some empirical results, *Journal of Financial Quantitative Analysis*, 10, 483-496.

Brennan, M.J., Schwartz, E.S. and Lagnado, R. (1997), Strategic asset allocation, *Journal of Economic Dynamics & Control*, 21, 1377-1403.

Brieman, L. (1960), Investment policies for expanding business optimal in a long run sense, *Naval Research Logistics Quarterly*, 7, 647-651.

Campbell J.Y. (1993), Intertemporal asset pricing without consumption data, *American Economic Review*, 83, 487-512.

Campbell, J.Y., Lo, A.W. and Mackinlay, A.C. (1997), *The Econometrics of Financial Markets*, Princeton University Press, Princeton.

Campbell, J.Y. and Mankiw, N.G. (1987), Are output fluctuations transitory? *Quarterly Journal of Economics*, 102, 857-880.

Carino, D.R., Kent, T., Myers, S.C., Sylvanus, M., Turner, A.L., Watanabe, K. and Ziemba, W.T. (1994), The Russell-Yasuda Kasai model: an asset/liability model for a Japanese insurance company using multistage stochastic programming, *Interfaces*, 24, 29-49.

Carino, D.R. and Ziemba, W.T. (1998), Formulation of the Russell-Yasuda Kasai financial planning model, *Operations Research*, 46, 433-449.

Chamberlain, G. (1983), A characterization of the distributions that imply mean-variance utility functions, *Journal of Economic Theory*, 29, 185-201.

Chan, L.C., Karceski, J. and Lakonishok, J. (1998), The risk and return from factors, *Journal of Financial and Quantitative Analysis*, 33, 159-188.

Chan, L.K., Yasushi, H. and Josef, L. (1991), Fundamentals and stock returns in Japan, *Journal of Finance*, 46, 1739-1789.

Chen, A.H.Y., Jen, F.C. and Zionts, S. (1971), The optimal portfolio revision policy, *Journal of Business*, 44, 51-61.

Chen, N.F. and Zhang, F. (1998), Risk and return of value stocks, *Journal of Business*, 71, 501-535.

Chen, S., Deng, X.T., Ma, C.Q. and Wang, S.Y. (2000), *Advances in Investment Decision Analysis*, Global-Link Publishers, Hong Kong.

Chen, S., Deng, X.T., Liu, W.G. and Wang, S.Y. (2000), Capital structure and portfolio selection, *Forecasting*, 11, 35-38.

Chen, S., Deng, X.T., Liu, W.G. and Wang, S.Y. (2001), Impact on the efficient frontier of portfolio of varying capital structure, *Chinese Journal of Management Science*, 9, 6-11.

Chen, S.N. (1986), An intertemporal capital asset pricing model under heterogeneous beliefs, *Journal of Economics and Business*, 38, 317-330.

Chou, A., Cooperstock, J., El-Yaniv, R., Klugerman, M. and Leighton, T. (1995), The statistical adversary allows optimal money-making trading strategies, in *Proceedings of the Sixth Annual ACM-SIAM Symposium on Discrete Algorithms*, 467-476.

Chunhachinda, P., Dandapani, K., Hamid, S. and Prakash, A.J. (1997), Portfolio selection and skewness: evidence from international stock markets, *Journal of Banking and Finance*, 21, 143-167.

Clarke, R.G., Krase, S. and Statman, M. (1994), Tracking errors, regret, and tactical asset allocation, *Journal of Portfolio Management*, 20, 16-24.

Cochrane, J.H. (1988), How big is the random walk in GNP? *Journal of Political Economy*, 96, 893-920.

Cohen, K. and Pogue, J. (1967), An empirical evaluation of alternative portfolio selection models, *Journal of Business*, 46, 166-193.

Conrad, J. and Kaul, G. (1988), Time-variation in expected returns, *Journal of Business*, 61, 409-425.

Constantinides, G.M. (1987), Theory of valuation: overview and recent developments, Working paper, University of Chicago.

Constantinides, G.M. and Duffie, D. (1996), Asset pricing with heterogeneous

consumers, *Journal of Political Economics*, 104, 219-240.
Cootner, B.B. (1989), *The Random Character of Stock Market Prices*, MIT Press, Cambridge.
Cover, T.M. (1991), Universal portfolios, *Mathematical Finance*, 1, 1-29.
Cover, T.M. and Ordentlich, E. (1996), Universal portfolios with side information, *IEEE Transactions on Information Theory*, 42, 348-363.
Cvitanic, J. and Karatzas, I. (1992), Convex duality in constrained portfolio optimization, *Annals of Applied Probability*, 2, 767-818.
Cvitanic, J. and Karatzas, I. (1993), Hedging contingent claims with constrained portfolios, *Annals of Applied Probability*, 3, 652-681.
Dantzig, G.B. and Infanger, G. (1993), Multi-stage stochastic linear programs for portfolio optimization, *Annals of Operations Research*, 45, 59-76.
Davis, M. H. A. and Norman, A. (1990), Portfolio selection with transaction costs, *Mathematics of Operations Research*, 15, 676-713.
Dembo, R.S. (1989), *Scenario Immunization*, Algorithmics Inc. Research Paper.
Dembo, R.S. (1990), Scenario optimization, *Annals of Operations Research*, 30, 63-80.
Dembo, R.D. and King, A.J. (1992), Tracking models and the optimal regret distribution, *Applied Stochastic Models and Data Analysis*, 8, 151-157.
Deng, X.T. (1996), Portfolio management with optimal regret ratio, in *Management Science and The Economic Development of China* (ed. by S. Ng *et al.*), HKUST, Hong Kong, 289-296.
Deng, X.T. (1999), Competitive ratio for portfolio management, in *Encyclopedia of Optimization* (ed. by C.A. Floudas, and P.M. Pardalos), Kluwer Academic Publishers, Boston.
Deng, X.T., Li, Z.F. and Wang, S.Y. (2000), On computation of arbitrage for markets with friction, in Computing and Combinatorics (ed. by D.Z. Du, P. Eades, V.Estivill-Castro, X. Lin and A. Sharma), Springer-Verlag, New York, 2000, 311-319.
Deng, X.T., Wang, S.Y. and Xia, Y.S. (2000), Criteria, models and strategies in portfolio selection, *Advanced Modeling and Optimization*, 2, 79-104.
Detemple, J. and Murthy, S. (1994), Intertemporal asset pricing with heterogeneous beliefs, *Journal of Economic Theory*, 62, 294-320.
Dimson, E. and Mussavian, M. (1999), Three centuries of asset pricing, *Journal of Bank and Finance*, 23, 1745-1769.
Dumas, B. and Luciano, E. (1991), An exact solution to a dynamic portfolio choice problem under transaction costs, *Journal of Finance*, 46, 577-595.
Dunn, K. and Singleton, K.J. (1986), Modeling the term structure of interest rates under non-separable utility and durable goods, *Journal of Financial Economics*, 17, 27-55.
Dybvig, P.H. and Ross, S.A. (1982), Yes, the APT is testable, *Journal of Finance*, 40, 1173-1188.
Elton, E.J. and Gruber, M.J. (1970), Homogeneous groups and the testing of

economic hypotheses, *Journal of Finance and Quantitative Analysis*, 9, 581-602.
Elton, E.J. and Gruber, M.J. (1971), Improved forecasting through design of homogeneous groups, *Journal of Business*, 44, 432-451.
Elton, E.J. and Gruber, M.J. (1973), Estimating the dependence structure of share prices - implications for portfolio selection, *Journal of Finance*, 28, 1203-1232.
Elton, E.J. and Gruber, M.J. (1974a), The multi-period consumption investment decision and single period analysis, *Oxford Economic Papers,* 26, 289-301.
Elton, E.J. and Gruber, M.J. (1974b), On the optimality of some multiperiod portfolio selection criteria, *Journal of Business*, 47, 231-243.
Elton, E.J. and Gruber, M.J. (1995), *Modern Portfolio Theory and Investment Analysis*, 5th Edition, John Wiley & Sons, New York.
Elton, E.J., Gruber, M.J. and Busse, J.A. (1998), Do investors care about sentiment? *Journal of Business*, 71, 477-500.
Elton, E.J., Gruber, M.J. and Urich, T.J. (1978), Are betas best? *Journal of Finance*, 33, 1357-1384.
El-Yaniv, R., Fiat, A., Karp, R.M. and Turpin, G. (1992), Competitive analysis of financial games, in *Proceedings of the 33rd Annual Symposium on Foundations of Computer Science*, 327-333.
Epstein, L.G. (1985), Decreasing risk aversion and mean-variance analysis, *Econometrica*, 53, 945-962.
Epstein, L.G. and Wang, T. (1994), Intertemporal asset pricing under Knightian uncertainty, *Econometrica*, 62, 283-322.
Fama, E.F. (1970), Multi-period consumption-investment decision, *American Economic Review*, 60, 163-174.
Fama, E.F. (1991), Efficient capital markets: II, *Journal of Finance*, XLVI, 1575-1617.
Fama, E.F. (1997), Multifactor portfolio efficiency and multifactor asset pricing, *Journal of Financial and Quantitative Analysis*, 32, 441-464.
Fama, E.F. and French, K.R. (1988), Permanent and temporary components of stock prices, *Journal of Political Economy*, 96, 246-273.
Fama, E.F. and French, K.R. (1992), The cross-section of expected stock returns, *Journal of Finance*, XLVII, 427-465.
Fama, E.F. and Macbeth, J.D. (1973), Risk, return and equilibrium: empirical tests, *Journal of Political Economy*, 81, 607-636.
Farrar, D.E. (1965), *The Investment Decision under Uncertainty,* Prentice-Hall, New York.
Farrell, J. (1974), Analyzing co-variation of returns to determine homogeneous stock grouping, *Journal of Business*, 47, 186-207.
Faust, J. (1992), When are variance ration tests for serial dependence optimal? *Econometrica*, 60, 1215-1226.
Fischer, D.E. and Jordan, R.J. (1987), *Security Analysis and Portfolio Management*, 4th Edition, Prentice-Hall, New York.
Fishburn, D.C. (1977), Mean-risk analysis with risk associated with below-target returns, *American Economical Review*, 67, 117-126.

Fleming, W.H. and Zariphopoulou, T. (1991), An optimal investment/consumption model with borrowing, *Mathematics of Operations Research*, 16, 802-822.

Gamrowski, B. and Rachev, S.T. (1999), A testable version of the Pareto-stable CAPM, *Mathematics and Computer Model*, 29, 61-81.

Gen, M. and Cheng, R.W. (1997), *Genetic Algorithms and Engineering Design,* John Wiley & Sons, New York.

Gennotte, G. and Jung, A. (1994), Investment strategies under transaction costs: the finite horizon case, *Management Science*, 40, 385-404.

Ghysels, E. and Hall, A. (1990), Are consumption-based intertemporal capital asset pricing models structural? *Journal of Econometrics*, 45, 121-139.

Gibbons, M.R. (1982), Multivariate tests of financial models: a new approach, *Journal of Financial Economics*, 10, 3-27.

Glover, F. and Jones, C.K. (1988), A stochastic network model and large scale mean-variance algorithm for portfolio selection, *Journal of Information and Optimization Sciences*, 9, 299-316.

Goldberg, D.E. (1989), *Genetic Algorithms in Search, Optimization and Machine Learning*, Addison-Wesley, New York.

Gonedes, N. (1976), Capital market equilibrium for a class of heterogeneous expectations in a two-parameter world, *Journal of Finance*, XXXI, 1-15.

Gonzalez-Gaverra, N. (1973), Inflation and capital asset market prices: theory and tests, *Ph.D. dissertation,* Stanford University, Stanford.

Green, R.C. (1986), Positively weighted portfolios on the minimum-variance frontier, *Journal of Finance*, 41, 1051-1068.

Grotschel, M., Lovasz, L. and Schrijver, A. (1987), *Geometric Algorithms and Combinatorial Optimization*, 2nd Edition, Springer-Verlag, Berlin.

Gultekin, M.N. and Gultekin, N.B. (1983), Stock market seasonality: international evidence, *Journal of Financial Economics*, 12, 469-481.

Hadar, J. and Russlell, W.R. (1969), Rules for ordering uncertain prospects, *American Economic Review*, 59, 25-34.

Hakansson, N. (1970), Optimal investment and consumption strategies under risk for a class of utility functions, *Econometrica,* 38, 587-607.

Hakansson, N. (1971), Multi-period mean-variance analysis: toward a general theory of portfolio choice, *Journal of Finance*, 26, 857-884.

Hakansson, N. (1974), Capital growth and the mean-variance approach to portfolio selection, *Journal of Financial and Quantitative Analysis*, VI, 517-557.

Hanoch, G. and Levy, H. (1969), The efficiency analysis of choices involving risk, *Review of Economic Studies*, 37, 335-346.

Hanoch, G. and Levy, H. (1970), Efficient portfolio selection with quadratic and cubic utility, *Journal of Business*, 52, 181-190.

Hansen, L.P. and Singleton, K.J. (1982), Generalized instrumental variables estimation of non-linear rational expectations models, *Econometrica*, 50, 1269-1286.

He, H. and Modest, D.M. (1995), Market frictions and consumption-based asset pricing, *Journal of Political Economics*, 103, 94-117.

Heaton, J. and Lucas, D.J. (1992), The effects of incomplete insurance markets and trading costs in a consumption-based asset pricing model, *Journal of Economic Dynamics & Control*, 16, 601-620.

Heaton, J. and Lucas, D.J. (1996), Evaluating the effects of incomplete markets on risk sharing and asset pricing, *Journal of Political Economics,* 104, 443-487.

Helmbold, D.P., Schapire, R.E., Singer, Y. and Warmuth, M.K. (1998), On-line portfolio selection using multiplicative updates, *Mathematical Finance*, 8, 325-347.

Hiller, R.S. and Eckstein, J. (1993), Stochastic dedication-designing fixed income portfolios using massively-parallel benders decomposition, *Management Science*, 39, 1422-1438.

Hiroshi, K. and Kenichi, S. (1995), A mean-variance-skewness optimization model, *Journal of the Operations Research Society of Japan*, 38, 173-187.

Hiroshi, K. and Hiroaki, Y. (1991), Mean-absolute deviation portfolio optimization model and its application to Tokyo stock market, *Management Science*, 37, 519-531.

Hiroshi, K., Stanley, R.P. and Suzuki, K.I. (1993), Optimal portfolios with asymptotic criteria, *Annals of Operations Research*, 45, 187-204.

Ho, M.S., Perraudin, W.R.M. and Sorensen, B.E. (1996), A continuous- time arbitrage-pricing model with stochastic volatility and jumps, *Journal of Business Economics and Statistics*, 14, 31-43.

Holland, J.H. (1975), *Adaptation in Natural and Artificial Systems*, University of Michigan Press, Ann Arbor.

Howison, S.D., Kelly, F.D. and Wilmott, S. (1995), *Mathematical Models in Finance*, Chapman & Hall, London.

Huang, C.F. and Litzenberger, R.H. (1988), *Foundations for Financial Economics*, Academic Press, New York.

Huberman, G., (1982), A simple approach to arbitrage pricing theory, *Journal of Economic Theory*, Vol. 28, 289-297.

Hull, J.C. (1993), *Options, Futures and Other Derivative Securities*, Prentice-Hall, New York.

Jacob, N.L. (1974), A limited-diversification portfolio selection model for the small investor, *Journal of Finance*, 29, 847-856.

Jagannathan, R., (1985), An investigation of commodity futures prices using the consumption-based intertemporal capital-asset pricing model, *Journal of Finance,* 40, 175-191.

Jegadeesh, N. (1991), Seasonality in stock price mean reversion: evidence form the U.S. and the U.K., *Journal of Finance*, XLVL, 1427-1444.

Jones, C.P. (1988), *Investments: Analysis and Management*, 2nd Edition, John Wiley & Sons, New York.

Kallberg, J.G. and Ziemba, W.T. (1983), Comparison of alternative utility functions in portfolio selection problems, *Management Science*, 29, 1257-1276.

Kandel, S. and Stambaugh, R.F. (1994), A mean-variance framework for tests of asset pricing models, *Review of Financial Studies*, 7, 803-804.

Karatzas, I., Lehoczky, J.P., Sethi, S.P. and Shreve, S.E., (1986), Explicit solution of a general consumption/investment problem, *Mathematics of Operations*

Research, 11, 261-294.

Karatzas, I., Lehochzky, J.P. and Shreve, S.E. (1987), Optimal portfolio and consumption decisions for a "smaller investor" on a finite horizon, *SIAM Journal on Control and Optimization*, 27, 1221-1259.

Karatzas, I., Lehoczky, J.P., Shreve, S.E. and Xu, G.L., (1991), Martingale and duality methods for utility maximization in an incomplete market, *SIAM Journal on Control and Optimization*, 29, 702-730.

Karoui, E. and Quenez, M.C. (1995), Dynamic programming and pricing of contingent claims in an incomplete market, *SIAM Journal on Control and Optimization*, 33, 29-66.

Keim, D.B. (1983), Size-related anomalies and stock return seasonality: further empirical evidence, *Journal of Financial Economics*, 12, 13-32.

King, B. (1966), Market and industry factors in stock price behavior, *Journal of Business*, 39, 139-140.

King, B. (1993), Asymmetric risk measures and tracking models for portfolio optimization under uncertainty, *Annals of Operations Research*, 45, 165-177.

Klaassen, P. (1998), Financial asset-pricing theory and stochastic programming models for asset/liability management: a synthesis, *Management Science*, 44, 31-48.

Klass, T. and Assaf, J. (1988), A diffusion model for optimal portfolio selection in the presence of brokerage fees, *Mathematics of Operations Research*, 13, 277-294.

Konno, H. and Suzuki, K. (1995), A mean-variance-skewness optimization model, *Journal of the Operations Research Society of Japan*, 38, 137-187.

Kroll, Y., Levy, H. and Markowitz, H. (1984), Mean-variance versus direct utility maximization, *Journal of Finance*, 39, 47-61.

Kumar, P.C., Philippatos, G.C. and Ezzell, J.R. (1978), Goal programming and the selection of portfolios by dual-purpose funds, *Journal of Finance*, 33, 303-310.

Kusy, M.I. and Ziemba, W.T., (1986), A bank asset and liability management model, *Operations Research*, 34, 356-376.

Lamont, O. (1998), Earnings and expected returns, *Journal of Finance*, LIII, 1563-1587.

Latane, H. (1959), Criteria for choice among risky ventures, *Journal of Political Economy*, 67, 144-155.

Latane, H. (1967), Criteria for portfolio building, *Journal of Finance*, 22, 359-373.

Latane, H. (1969), Test of portfolio building rules, *Journal of Finance*, 24, 595-612.

Lee, S.M. and Lerro, A.J. (1973), Optimizing the portfolio for mutual funds, *Journal of Finance*, 28, 1089-1102.

Lehoczky, J., Sethi, S. and Shreve, S. (1983), Optimal consumption and investment policies allowing consumption constraints and bankruptcy, *Mathematics of Operations Research*, 8, 613-636.

Levy, H. (1978), Equilibrium in an imperfect market: a constraint on the number of securities in the portfolio, *American Economic Review*, 68, 643-658.

Levy, H. and Markowitz, H. M. (1979), Approximating expected utility by a function of mean and variance, *American Economic Review*, 69, 308-317.

Levy, H. and Samuelson, P. A. (1992), The capital asset pricing model with diverse holding periods, *Management Science*, 38, 1529-1542.

Levy, H. and Sarnat, M. (1970), Alternative efficiency criteria: an empirical analysis, *Journal of Finance*, 25, 1153-1158.

Lewis, K.K. (1991), Should the holding period matter for the intertemporal consumption-based CAPM, *Journal of Monetary Economics*, 28, 365-389.

Li, D. and Ng, W.L. (2000), Optimal dynamic portfolio selection: multiperiod mean variance formulation, *Mathematical Finance*, 10, 387-406.

Li, Z.F., Li, Z.X. and Wang, S.Y. (2000a), An analytic derivation of multi-period efficient portfolio frontier, submitted to *Journal of Finance.*

Li, Z.F., Li, Z.X. and Wang, S.Y. (2000b), An explicit expression of efficient portfolio frontier of un-correlated assets with no short sales, submitted to *European Journal of Operational Research.*

Li, Z.F., Li, Z.X., Wang, S.Y. and Deng, X.T. (2001), Optimal portfolio selection of assets with transaction costs and no short sales, *International Journal of Systems Science*, 32, 599-607.

Li, Z.F. and Wang, S.Y. (2001), *Portfolio Optimization and No-arbitrage (in Chinese)*, Science Press, Beijing.

Li, Z.F., Wang, S.Y. and Deng, X.T. (2000), A linear programming algorithm for optimal portfolio selection with transaction costs, *International Journal of Systems Science,* 31, 107-117.

Li, Z.F., Wang, S.Y. and Yang, H.L. (2000), Characterization of weak no-arbitrage in frictional markets, submitted to *Mathematical Finance.*

Lindenberg, E. (1976), Imperfect competition among investors in security markets, *Ph.D. Dissertation,* New York University, New York.

Lindenberg, E. (1979), Capital market equilibrium with price affecting institutional investors, in *Portfolio Theory 25 Years Later* (ed. by Elton and Gruber), North-Holland, Amsterdam.

Lintner, J. (1965), The valuation of risk assets and the selection of risky investments in stock portfolios and capital budgets, *Review of Economic Statistics*, 47, 13-37.

Lintner, J. (1969), The aggregation of investor diverse judgments and preferences in purely competitive security markets, *Journal of Financial and Quantitative Analysis*, 4, 347-400.

Litzenberger, R.H. and Ramaswamy, K. (1979), The effect of personal taxes and dividends on capital asset prices: theory and empirical evidence, *Journal of Financial Economics*, 7, 163-195.

Liu, S.C., Wang, S.Y. and Qiu, W.H. (2000a), A universal portfolio with transaction costs, in *Advances in Investment Decision Analysis* (ed. by S. Chen *et al.*), Global-Link Publishers, Hong Kong, 151-162.

Liu, S.C., Wang, S.Y. and Qiu, W.H. (2000b), An approach for portfolio selection based on entropy, *Systems Engineering: Theory and Practice*, 18, 245-253.

Liu, S.C., Wang, S.Y. and Qiu, W.H. (2000c), A mean-variance-skewness model for portfolio selection with transaction costs, submitted to *International Journal of Systems Sciences.*

Liu, S.C., Wang, S.Y. and Qiu,W.H. (2001), A game model for portfolio selection, *Systems Engineering: Theory and Practice*, 19, 56-63.

Loeb, T.F. (1983), Trading costs: the critical link between investment information and results, *Financial Analysts Journal*, May/June, 39-44.

Lucas, D.J. (1994), Asset pricing with un-diversifiable income risk and short sales constraints-deepening the equity premium puzzle, *Journal of Monetary Economics*, 34, 325-341.

Luttmer, E.G.J. (1996), Asset pricing in economies with frictions, *Econometrica,* 64, 1439-1467.

Maclean, L.C. and Ziemba, W.T. (1991), Growth-security profiles in capital accumulation under risk, *Annals of Operations Research*, 31, 501-510.

Maclean, L.C., Ziemba, W.T. and Blazenko, G. (1992), Growth versus security in dynamic investment analysis, *Management Science*, 38, 1562-1585.

Magill, M. and Quinzii, M. (2000), Infinite horizon CAPM equilibrium, *Economic Theory*, 15, 103-138.

Malliaris, A.G. and Stein, J.L. (1999), Methodological issues in asset pricing: Random walk or chaotic dynamics, *Journal of Bank and Finance*, 23, 1605-1635.

Mandelbrotm, B.B. (1989), *In the New Palgrave, A Dictionary of Economics*, Maxmillan Press Limited, London.

Mao, J.C.T. (1970a), Survey of capital budgeting: theory and practice, *Journal of Finance*, 25, 349-360.

Mao, J.C.T. (1970b), Models of capital budgeting, E-V vs. E-S, *Journal of Financial and Quantitative Analysis*, 5, 657-675.

Mao, J.C.T. (1970c), Essentials of portfolio diversification strategy, *Journal of Finance*, 25, 1109-1121.

Markowitz, H. (1952), Portfolio selection, *Journal of Finance*, 7, 77-91.

Markowitz, H. (1959), *Portfolio Selection, Efficient Diversification of Investment*, Basil Blackwell Ltd., New York.

Markowitz, H. (1976), Investment for the long-run: new evidence for an old rule, *Journal of Finance*, 31, 1273-1286.

Markowitz, H. (1987), *Mean-Variance Analysis in Portfolio Choice and Capital Markets*, Blackwell Publishers, Cambridge.

Mcentire, P.L. (1984), Portfolio theory for independent assets, *Management Science*, 30, 952-963.

Merton, R.C. (1969), Lifetime portfolio selection under uncertainty: the continuous case, *Review of Economics and Statistics*, 51, 247-257.

Merton, R.C. (1971), Optimum consumption and portfolio rules in a continuous time model, *Journal of Economic Theory*, 3, 373-413.

Merton, R.C. (1972), An analytical derivation of the efficient portfolio frontier, *Journal of Financial and Quantitative Analysis*, 7, 1851-1872.

Merton, R.C. (1973a), An intertemporal capital asset pricing model, *Econometrica*, 41, 867-887.

Merton, R.C. (1973b), Theory of rational option pricing, *Bell Journal of Economics and Management Science*, 4, 141-183.

Merton, R.C. (1990), *Continuous-Time Finance*, Blackwell Publishers, Cambridge, MA.

Merton, R.C. and Samuelson, P.A. (1974), Fallacy of the log-normal approximation to optimal portfolio decision-making over many periods, *Journal of Financial Economics*, 1, 67-94.

Michalewicz, Z. (1994), *Genetic Algorithms + Data Structures = Evolution Programs*, Springer-Verlag, New York.

Michaud, R.O. (1981), Risk policy and long-term investment, *Journal of Financial and Quantitative Analysis*, 16, 147-167.

Miller, M.H. and Scholes, M. (1972), Rates of return in relation to risk: a re-examination of some recent findings, in *Studies in the Theory of Capital Markets* (ed. by S. Jensen), Praeger, New York.

Morton, A.J. and Pliska, S. R. (1995), Optimal portfolio management with transaction costs, *Mathematical Finance*, 5, 337-356.

Morton, D. (2000), *Stochastic Optimization: Class Notes*, University of Texas at Austin, Austin.

Mossin, J. (1966), Equilibrium in a capital asset market, *Econometrica*, 34, 768-783.

Mossin, J. (1968), Optimal multiperiod portfolio policies, *Journal of Business*, 41, 215-229.

Mulvey, J.M., Gould, G. and Morgan, C. (2000), An asset and liability management system for Towers Perrin-Tillinghast, *Interfaces*, 30, 96-114.

Mulvey, J.M., Rosenbaum, D.P. and Shetty, B. (1997), Strategic financial risk management and operations research, *European Journal of Operational Research*, 97, 1-16.

Mulvey, J. M. and Vladimirou, H. (1992), Stochastic network programming for financial planning problems, *Management Science*, 38, 1642-1664.

Munk, C. (1997), No-arbitrage bounds on contingent claims prices with convex constraints on the dollar investments of the hedge portfolio, Working paper, Odense University, Odense.

Nichols, W.D. and Brown, S.L. (1981), Assimilating earnings and split information: is the capital market becoming more efficient? *Journal of Financial Economics*, 9, 309-315.

Oh, G.T. (1996), Some results in the CAPM with non-traded endowments, *Management Science*, 42, 286-293.

Ohlson, J.A. and Garman, M.B. (1980), A dynamic equilibrium for the Ross arbitrage model, *Journal of Finance*, 35, 675-684.

Ordentlich, E. and Cover, T.M. (1998), The cost of achieving the best portfolio in hindsight, *Mathematics of Operations Research*, 23, 960-982.

Ostermark, R. (1991), Vector forecasting and dynamic portfolio selection: empirical efficiency of recursive multiperiod strategies, *European Journal of*

Operational Research, 55, 45-56.

Pardalos, P.M., Sandstrom, M. and Zopounidis, C. (1994), On the use of optimization models for portfolio selection: a review and some computational results, *Computational Economics*, 7, 227-244.

Pastor, L. (2000), Portfolio selection and asset pricing models, *Journal of Finance*, 55, 179-223.

Patel, N.R. and Subrahmanyam, M.G. (1982), A simple algorithm for optimal portfolio selection with fixed transaction costs, *Management Science*, 28, 303-314.

Perold, A.F. (1984), Large-scale portfolio optimization, *Management Science*, 30, 1143-1160.

Perold A.F. and Sharpe, W.F. (1988), Dynamic strategies for asset allocations, *Financial Analysts Journal*, Jan/Feb, 16-27.

Philp, H., Stephen, D. and Ross, A. (1982), Portfolio efficient, *Econometric*, 50, 1525-1546.

Philippatos, G.C. and Gressis, N. (1975), Conditions of formal equivalence among E-V, SSD, and E-H portfolio selection criteria: the case for uniform, normal, and lognormal distributions, *Management Science*, 11, 617-625.

Philippatos, G.C. and Wilson, C.J. (1972), Entropy, market risk, and the selection of efficient portfolios, *Applied Economics*, 4, 209-220.

Philippatos, G.C. and Wilson, C.J. (1974), Entropy, market risk, and the selection of efficient portfolios: reply, *Applied Economics*, 6, 77-81.

Pliska, S. (1986), A stochastic calculus model of continuous trading: optimal portfolios, *Mathematics of Operations Research*, 11, 371-382.

Pliska, S.R. (1997), *Introduction to Mathematical Finance*, Blackwell Publishers Inc., New York.

Pogue, J.A. (1970), An extension of the Markowitz portfolio selection model to include variable transactions costs, short sales, leverage policies, and taxes, *Journal of Finance*, 25, 1005-1028.

Porter, R.B. (1973), An empirical comparison of stochastic dominance and mean variance portfolio choice criteria, *Journal of Finance and Quantitative Analysis*, 8, 587-608.

Quirk, J. and Saposnik, R. (1962), Admissibility and measurable utility functions, *Review of Economic Studies*, 30, 140-146.

Ramchand, L. (1999), Asset pricing in open economies with incomplete markets: implications for foreign currency returns, *Journal of International Money Finance*, 18, 871-890.

Reisman, H. (1992a), Reference variables, factors structure, and the approximate multi-beta representation, *Journal of Finance*, 47, 1303-1314.

Reisman, H. (1992b), Intertemporal arbitrage pricing theory, *Review of Financial Studies*, 5, 105-122.

Richardson, M. (1993), Temporary components of stock prices: a skeptic's view, *Journal of Business & Economic Statistics*, 11, 199-207.

Richardson, M. and Smith, T. (1991), Tests of financial models in the presence of overlapping observations, *Review of Financial Studies*, 4, 227-254.

Robinson, S. M. (1991), Extended scenario analysis, *Annals of Operations Research*, 31, 385-398.

Rockafellar, R.T. (1970), *Convex Analysis*, Princeton University Press, Princeton.

Rockafellar, R.T. and Wets, R.J.-B. (1991), Scenario and policy aggregation in optimization under uncertainty, *Mathematics of Operations Research*, 16, 119-147.

Roll, R. (1973), Evidence on the ``growth optimum'' model, *Journal of Finance*, 28, 551-556.

Roll, R. (1977), A critique of the asset pricing theory's tests, *Journal of Financial Economics*, 4, 129-176.

Roll, R. (1992), A mean/variance analysis of tracking error, *Journal of Portfolio Management*, 19, 13-22.

Roll, R. and Ross, S.A., (1980) A empirical investigation of the arbitrage pricing theory, *Journal of Finance*, 35, 1073–1103.

Rosenberg, B., Kenneth, R., and Ronald, L. (1985), Persuasive evidence of market inefficiency, *Journal of Portfolio Management*, 11, 9-17.

Ross, S.A. (1976), The arbitrage theory of capital asset pricing, *Journal of Economic Theory*, 13, 341-360.

Ross, S.A. (1977), Risk, return and arbitrage, in *Risk and Returns in Finance* (ed. by I. Friend and J.L. Bicksler), MIT Press, Cambridge, 189-218.

Rothschild, M. and Stiglitz, J. (1970), Increasing risk: I, a definition, *Journal of Economic Theory*, 2, 225-243.

Rothschild, M and Stiglitz, J. (1971), Increasing risk: II, its economic consequences, *Journal of Economic Theory*, 3, 66-84.

Roy, A.D. (1952), Safety-first and the holding of assets, *Econometrics*, 20, 431-449.

Rozeff, M.S. (1984), Dividend yields are equity risk premiums, *Journal of Portfolio Management*, 11, 68-75.

Rubinstein, M. (1976), The valuation of uncertain income streams and the pricing of options, *Bell Journal of Economics and Management Science*, 7, 407-425.

Russell, W.R. and Smith, P.E. (1966), Communications to editor: a comment on Baumol efficient portfolios, *Management Science*, 12, 619-621.

Samuelson, P. (1958), The fundamental approximation theorem of portfolio analysis in terms of means variances and higher moments, *Review of Economic Studies*, 25, 65-86.

Samuleson, P. (1969), Lifetime portfolio selection by dynamic stochastic programming, *Review of Economics and Statistics*, 51, 239-246.

Samuelson, P. (1997), How best to flip-flop if you must: integer dynamic stochastic programming for either-or, *Journal of Risk and Uncertainty*, 15, 183-190.

Schreiner, J. and Smith, K. (1980), The impact of mayday on diversification costs, *Journal of Portfolio Management*, 6, 28-36.

Schultz, P. (1983), Transaction costs and the small firm effect: a comment, *Journal of Financial Economics*, 12, 81-88.

Sengupta, J.K. (1983), Optimal portfolio investment in a dynamic horizon,

International Journal of Systems Science, 14, 789-800.

Sengupta, J.K. (1986), *Stochastic Optimization and Economic Models*, D. Reidal Publishing Company, New York.

Sengupta, J.K. (1989a), Portfolio decisions as games, *International Journal of Systems Science*, 20, 1323-1334.

Sengupta, J.K. (1989b), Mixed strategy and information theory in optimal portfolio choice, *International Journal of Systems Science*, 20, 215-227.

Seshadri, S., Khanna, A. and Harche, F. (1999), A method for strategic asset-liability management with an application to the federal home loan bank of New York, *Operations Research*, 47, 345-360.

Shanken, J. (1982), The arbitrage pricing theory: is it testable? *Journal of Finance*, 37, 1129-1140.

Shanken, J. (1985), Multi-beta CAPM or equilibrium APT? a reply. *Journal of Finance*, 40, 1189-1196.

Shanken, J. (1987), Multivariate proxies and asset pricing relations: living with the roll critique, *Journal of Financial Economics*, 18, 91-110.

Shanken, J. (1992), The current state of the arbitrage pricing theory, *Journal of finance*, 47, 1569–1574.

Sharpe, W.F. (1963), A simplified model for portfolio analysis, *Management Science*, 9, 277-293.

Sharpe, W.F. (1964), Capital asset prices: a theory of market equilibrium under conditions of risk, *Journal of Finance*, 19, 425-442.

Sharpe, W.F. (1967), A linear programming algorithm for mutual funds portfolio selection, *Management Science*, 13, 499-510.

Sharpe, W.F. (1998), www.stanford.edu/~wfsharpe/art/djam/djam.htm

Sharpe, W.F. and Cooper, G.M. (1992), Risk-return class of New York Stock Exchange common stocks: 1931-1967, *Financial Analysts Journal*, 28, 46-52.

Shreve, S.E. and Soner, H.M. (1994), Optimal investment and consumption with transaction costs, *Annals of Applied Probabilities*, 4, 609-692.

Simaan, Y. (1997), Estimation risk in portfolio selection: the mean variance model versus the mean absolute deviation model, *Management Science*, 43, 1437-1446.

Smith, K.V. (1967), A transition model for portfolio revision, *Journal of Finance*, 22, 425-439.

Stambaugh, R.F. (1982), On the exclusion of assets from tests of the two-parameter model, *Journal of Financial Economics*, 10, 237-268.

Statman, M. (1987), How many stocks make a diversified portfolio? *Journal of Financial and Quantitative Analysis*, 22, 353-363.

Stattman, D. (1980), Book values and stock returns, *The Chicago MBA: A Journal of Selected Papers,* 4, 25-45.

Swalm, R.O. (1966), Utility theory-insights into risk taking, *Harvard Business Review*, 44, 123-136.

Tamiz, M. (1996), *Multi-Objective Programming and Goal Programming*, Lecture Notes in Economic and Mathematical Systems, 432, Springer, Berlin.

Tucker, A.L. (1992), *Financial Futures, Options and Swaps*, West Publishing

Company, New York.

Turtle, H.J. (1994), Temporal dependence in asset pricing-models, *Economics Letters*, 45, 361-366.

Vasicek, O. (1973), A note on using cross-sectional information in Bayesian estimation of security Betas, *Journal of Finance*, 8, 1233-1239.

Wang, J. (1993), A model of intertemporal asset prices under asymmetric information, *Review of Economic Study*, 60, 249-282.

Wang, S.Y. (2000), On competitive solutions to online financial problems, in *Proceedings of International Symposium on Knowledge and Systems Sciences: Challenges to Complexity* (ed. by Y.Nakamori, J.F.Gu and T.Yoshida), JAIST, Ishikawa, 162-168.

Wang, X.J. (1995), Equilibrium-models of asset pricing with progressive taxation and tax evasion, *Applied Economics Letters*, 2, 440-443.

Whitemore, G.A. (1970), Third-degree stochastic dominance, *American Economic Review*, 60, 457-459.

Wilcox, J.W. (1998), Investing at the edge, *Journal of Portfolio Management*, 24, 9-21.

Williams, J.O. (1997), Maximizing the probability of achieving investment goals, *Journal of Portfolio Management*, 46, 77-81.

Winkler R.L. and Barry C.B. (1975), A Bayesian model for portfolio selection and revision, *Journal of Finance*, 30, 179-192.

Xia, Y.S., Liu, B.D., Wang, S.Y. and Lai, K.K. (2000), A model for portfolio selection with order of expected returns, *Computers & Operations Research,* 27, 409-422.

Xia, Y.S., Wang, S.Y. and Deng, X.T. (1998a), An optimal portfolio selection model with transaction costs, in *Advances in Operations Research and Systems Engineering* (edited by J.F. Gu *et al.*), Global-Link Publishers, Hong Kong, 118-123.

Xia, Y.S., Wang, S.Y. and Deng, X.T. (1998b), Mathematical models in finance, *Chinese Journal of Management Science*, 6, 1-13.

Xia, Y.S., Wang, S.Y. and Deng, X.T. (1999), Empirical tests of CAPM for the stock markets of China, submitted to *Journal of Business*.

Xia, Y.S., Wang, S.Y. and Deng, X.T. (2001), A compromise solution to mutual fund portfolios with transaction costs, *European Journal of Operations Research*, 134, 564-581.

Yoshimoto, A. (1996), The Mean-variance approach to portfolio optimization subject to transaction costs, *Journal of the Operations Research Society of Japan*, 39, 99-117.

Young, W.E. and Trent, R.H. (1969), Geometric mean approximation of individual security and portfolio management, *Journal of Financial Quantitative Analysis*, 4, 179-199.

Yu, P.L. (1985), *Multiple Criteria Decision Making: Concepts, Techniques and Extensions*, Plenum Press, New York.

Zariphopoulou, T. (1992), Investment-consumption models with transaction fees and Markov chain parameters, *SIAM Journal on Control and Optimization*, 30, 613–636.

Zariphopoulou, T. (1994), Consumption-investment models with constraints, *SIAM Journal of Control and Optimization*, 32, 59-85.

Zariphopoulou, T. (1997), Optimal investment and consumption models with non-linear stock dynamics, *Mathematical Method of Operations Research*, 50, 271-296.

Zenios, S.A. (1991), *Financial Optimization*, Cambridge University Press, New York.

Zenios, S.A., Holmer, M.R., Mckendall, R. and Vassiadou-Zeniou, C. (1998), Dynamic models for fixed-income portfolio management under uncertainty, *Journal of Economic Dynamics & Control*, 22, 1517-1541.

Zhou, C.S. (1998), Dynamic portfolio choice and asset pricing with differential information, *Journal of Economic Dynamics & Control*, 22, 1027-1051.

Zhou, X.Y. and Li, D. (1999), Explicit efficient frontier of a continuous time mean variance portfolio selection problem: in *Control of Distributed Parameter and Stochastic Systems*, Kluwer, Boston, 323-340.

Zhou, Y., Chen, S., Deng, X.T. and Wang, S.Y. (2000), Influence of varying interest rates on efficient frontier, *Chinese Journal of Management Science*, 14, 828-835.

Zhou, Y., Chen, S. and Wang, S.Y. (2000), On the trail of the optimal portfolio selection under the influence of the capital structure, *Journal of Management Sciences*, 3, 245-253.

Ziemba, W.T., Parkan, C. and Brooks-Hill, F.J. (1974), Calculation of investment portfolios with risk free borrowing and lending, *Management Science*, 21, 209-222.

Ziemba, W.T. and Vickson, R.G. (1975), *Stochastic Optimization Models in Finance*, Academic Press, New York.

Subject Index

Author Index

Lecture Notes in Economics and Mathematical Systems

For information about Vols. 1–320
please contact your bookseller or Springer-Verlag

Vol. 325: P. Ferri, E. Greenberg, The Labor Market and Business Cycle Theories. X, 183 pages. 1989.

Vol. 326: Ch. Sauer, Alternative Theories of Output, Unemployment, and Inflation in Germany: 1960–1985. XIII, 206 pages. 1989.

Vol. 327: M. Tawada, Production Structure and International Trade. V, 132 pages. 1989.

Vol. 328: W. Güth, B. Kalkofen, Unique Solutions for Strategic Games. VII, 200 pages. 1989.

Vol. 329: G. Tillmann, Equity, Incentives, and Taxation. VI, 132 pages. 1989.

Vol. 330: P.M. Kort, Optimal Dynamic Investment Policies of a Value Maximizing Firm. VII, 185 pages. 1989.

Vol. 331: A. Lewandowski, A.P. Wierzbicki (Eds.), Aspiration Based Decision Support Systems. X, 400 pages. 1989.

Vol. 332: T.R. Gulledge, Jr., L.A. Litteral (Eds.), Cost Analysis Applications of Economics and Operations Research. Proceedings. VII, 422 pages. 1989.

Vol. 333: N. Dellaert, Production to Order. VII, 158 pages. 1989.

Vol. 334: H.-W. Lorenz, Nonlinear Dynamical Economics and Chaotic Motion. XI, 248 pages. 1989.

Vol. 335: A.G. Lockett, G. Islei (Eds.), Improving Decision Making in Organisations. Proceedings. IX, 606 pages. 1989.

Vol. 336: T. Puu, Nonlinear Economic Dynamics. VII, 119 pages. 1989.

Vol. 337: A. Lewandowski, I. Stanchev (Eds.), Methodology and Software for Interactive Decision Support. VIII, 309 pages. 1989.

Vol. 338: J.K. Ho, R.P. Sundarraj, DECOMP: An Implementation of Dantzig-Wolfe Decomposition for Linear Programming. VI, 206 pages.

Vol. 339: J. Terceiro Lomba, Estimation of Dynamic Econometric Models with Errors in Variables. VIII, 116 pages. 1990.

Vol. 340: T. Vasko, R. Ayres, L. Fontvieille (Eds.), Life Cycles and Long Waves. XIV, 293 pages. 1990.

Vol. 341: G.R. Uhlich, Descriptive Theories of Bargaining. IX, 165 pages. 1990.

Vol. 342: K. Okuguchi, F. Szidarovszky, The Theory of Oligopoly with Multi-Product Firms. V, 167 pages. 1990.

Vol. 343: C. Chiarella, The Elements of a Nonlinear Theory of Economic Dynamics. IX, 149 pages. 1990.

Vol. 344: K. Neumann, Stochastic Project Networks. XI, 237 pages. 1990.

Vol. 345: A. Cambini, E. Castagnoli, L. Martein, P Mazzoleni, S. Schaible (Eds.), Generalized Convexity and Fractional Programming with Economic Applications. Proceedings, 1988. VII, 361 pages. 1990.

Vol. 346: R. von Randow (Ed.), Integer Programming and Related Areas. A Classified Bibliography 1984–1987. XIII, 514 pages. 1990.

Vol. 347: D. Ríos Insua, Sensitivity Analysis in Multi-objective Decision Making. XI, 193 pages. 1990.

Vol. 348: H. Störmer, Binary Functions and their Applications. VIII, 151 pages. 1990.

Vol. 349: G.A. Pfann, Dynamic Modelling of Stochastic Demand for Manufacturing Employment. VI, 158 pages. 1990.

Vol. 350: W.-B. Zhang, Economic Dynamics. X, 232 pages. 1990.

Vol. 351: A. Lewandowski, V. Volkovich (Eds.), Multiobjective Problems of Mathematical Programming. Proceedings, 1988. VII, 315 pages. 1991.

Vol. 352: O. van Hilten, Optimal Firm Behaviour in the Context of Technological Progress and a Business Cycle. XII, 229 pages. 1991.

Vol. 353: G. Ricci (Ed.), Decision Processes in Economics. Proceedings, 1989. III, 209 pages 1991.

Vol. 354: M. Ivaldi, A Structural Analysis of Expectation Formation. XII, 230 pages. 1991.

Vol. 355: M. Salomon. Deterministic Lotsizing Models for Production Planning. VII, 158 pages. 1991.

Vol. 356: P. Korhonen, A. Lewandowski, J . Wallenius (Eds.), Multiple Criteria Decision Support. Proceedings, 1989. XII, 393 pages. 1991.

Vol. 357: P. Zörnig, Degeneracy Graphs and Simplex Cycling. XV, 194 pages. 1991.

Vol. 358: P. Knottnerus, Linear Models with Correlated Disturbances. VIII, 196 pages. 1991.

Vol. 359: E. de Jong, Exchange Rate Determination and Optimal Economic Policy Under Various Exchange Rate Regimes. VII, 270 pages. 1991.

Vol. 360: P. Stalder, Regime Translations, Spillovers and Buffer Stocks. VI, 193 pages . 1991.

Vol. 361: C. F. Daganzo, Logistics Systems Analysis. X, 321 pages. 1991.

Vol. 362: F. Gehrels, Essays in Macroeconomics of an Open Economy. VII, 183 pages. 1991.

Vol. 363: C. Puppe, Distorted Probabilities and Choice under Risk. VIII, 100 pages . 1991

Vol. 364: B. Horvath, Are Policy Variables Exogenous? XII, 162 pages. 1991.

Vol. 365: G. A. Heuer, U. Leopold-Wildburger. Balanced Silverman Games on General Discrete Sets. V, 140 pages. 1991.

Vol. 366: J. Gruber (Ed.), Econometric Decision Models. Proceedings, 1989. VIII, 636 pages. 1991.

Vol. 367: M. Grauer, D. B. Pressmar (Eds.), Parallel Computing and Mathematical Optimization. Proceedings. V, 208 pages. 1991.

Vol. 368: M. Fedrizzi, J. Kacprzyk, M. Roubens (Eds.), Interactive Fuzzy Optimization. VII, 216 pages. 1991.

Vol. 369: R. Koblo, The Visible Hand. VIII, 131 pages.1991.

Vol. 370: M. J. Beckmann, M. N. Gopalan, R. Subramanian (Eds.), Stochastic Processes and their Applications. Proceedings, 1990. XLI, 292 pages. 1991.

Vol. 371: A. Schmutzler, Flexibility and Adjustment to Information in Sequential Decision Problems. VIII, 198 pages. 1991.

Vol. 372: J. Esteban, The Social Viability of Money. X, 202 pages. 1991.

Vol. 373. A. Billot, Economic Theory of Fuzzy Equilibria. XIII, 164 pages. 1992.

Vol. 374: G. Pflug, U. Dieter (Eds.), Simulation and Optimization. Proceedings, 1990. X, 162 pages. 1992.

Vol. 375: S.-J. Chen, Ch.-L. Hwang, Fuzzy Multiple Attribute Decision Making. XII, 536 pages. 1992.

Vol. 376: K.-H. Jöckel, G. Rothe, W. Sendler (Eds.), Bootstrapping and Related Techniques. Proceedings, 1990. VIII, 247 pages. 1992.

Vol. 377: A. Villar, Operator Theorems with Applications to Distributive Problems and Equilibrium Models. XVI, 160 pages. 1992.

Vol. 378: W. Krabs, J. Zowe (Eds.), Modern Methods of Optimization. Proceedings, 1990. VIII, 348 pages. 1992.

Vol. 379: K. Marti (Ed.), Stochastic Optimization. Proceedings, 1990. VII, 182 pages. 1992.

Vol. 380: J. Odelstad, Invariance and Structural Dependence. XII, 245 pages. 1992.

Vol. 381: C. Giannini, Topics in Structural VAR Econometrics. XI, 131 pages. 1992.

Vol. 382: W. Oettli, D. Pallaschke (Eds.), Advances in Optimization. Proceedings, 1991. X, 527 pages. 1992.

Vol. 383: J. Vartiainen, Capital Accumulation in a Corporatist Economy. VII, 177 pages. 1992.

Vol. 384: A. Martina, Lectures on the Economic Theory of Taxation. XII, 313 pages. 1992.

Vol. 385: J. Gardeazabal, M. Regúlez, The Monetary Model of Exchange Rates and Cointegration. X, 194 pages. 1992.

Vol. 386: M. Desrochers, J.-M. Rousseau (Eds.), Computer-Aided Transit Scheduling. Proceedings, 1990. XIII, 432 pages. 1992.

Vol. 387: W. Gaertner, M. Klemisch-Ahlert, Social Choice and Bargaining Perspectives on Distributive Justice. VIII, 131 pages. 1992.

Vol. 388: D. Bartmann, M. J. Beckmann, Inventory Control. XV, 252 pages. 1992.

Vol. 389: B. Dutta, D. Mookherjee, T. Parthasarathy, T. Raghavan, D. Ray, S. Tijs (Eds.), Game Theory and Economic Applications. Proceedings, 1990. IX, 454 pages. 1992.

Vol. 390: G. Sorger, Minimum Impatience Theorem for Recursive Economic Models. X, 162 pages. 1992.

Vol. 391: C. Keser, Experimental Duopoly Markets with Demand Inertia. X, 150 pages. 1992.

Vol. 392: K. Frauendorfer, Stochastic Two-Stage Programming. VIII, 228 pages. 1992.

Vol. 393: B. Lucke, Price Stabilization on World Agricultural Markets. XI, 274 pages. 1992.

Vol. 394: Y.-J. Lai, C.-L. Hwang, Fuzzy Mathematical Programming. XIII, 301 pages. 1992.

Vol. 395: G. Haag, U. Mueller, K. G. Troitzsch (Eds.), Economic Evolution and Demographic Change. XVI, 409 pages. 1992.

Vol. 396: R. V. V. Vidal (Ed.), Applied Simulated Annealing. VIII, 358 pages. 1992.

Vol. 397: J. Wessels, A. P. Wierzbicki (Eds.), User-Oriented Methodology and Techniques of Decision Analysis and Support. Proceedings, 1991. XII, 295 pages. 1993.

Vol. 398: J.-P. Urbain, Exogeneity in Error Correction Models. XI, 189 pages. 1993.

Vol. 399: F. Gori, L. Geronazzo, M. Galeotti (Eds.), Nonlinear Dynamics in Economics and Social Sciences. Proceedings, 1991. VIII, 367 pages. 1993.

Vol. 400: H. Tanizaki, Nonlinear Filters. XII, 203 pages. 1993.

Vol. 401: K. Mosler, M. Scarsini, Stochastic Orders and Applications. V, 379 pages. 1993.

Vol. 402: A. van den Elzen, Adjustment Processes for Exchange Economies and Noncooperative Games. VII, 146 pages. 1993.

Vol. 403: G. Brennscheidt, Predictive Behavior. VI, 227 pages. 1993.

Vol. 404: Y.-J. Lai, Ch.-L. Hwang, Fuzzy Multiple Objective Decision Making. XIV, 475 pages. 1994.

Vol. 405: S. Komlósi, T. Rapcsák, S. Schaible (Eds.), Generalized Convexity. Proceedings, 1992. VIII, 404 pages. 1994.

Vol. 406: N. M. Hung, N. V. Quyen, Dynamic Timing Decisions Under Uncertainty. X, 194 pages. 1994.

Vol. 407: M. Ooms, Empirical Vector Autoregressive Modeling. XIII, 380 pages. 1994.

Vol. 408: K. Haase, Lotsizing and Scheduling for Production Planning. VIII, 118 pages. 1994.

Vol. 409: A. Sprecher, Resource-Constrained Project Scheduling. XII, 142 pages. 1994.

Vol. 410: R. Winkelmann, Count Data Models. XI, 213 pages. 1994.

Vol. 411: S. Dauzère-Péres, J.-B. Lasserre, An Integrated Approach in Production Planning and Scheduling. XVI, 137 pages. 1994.

Vol. 412: B. Kuon, Two-Person Bargaining Experiments with Incomplete Information. IX, 293 pages. 1994.

Vol. 413: R. Fiorito (Ed.), Inventory, Business Cycles and Monetary Transmission. VI, 287 pages. 1994.

Vol. 414: Y. Crama, A. Oerlemans, F. Spieksma, Production Planning in Automated Manufacturing. X, 210 pages. 1994.

Vol. 415: P. C. Nicola, Imperfect General Equilibrium. XI, 167 pages. 1994.

Vol. 416: H. S. J. Cesar, Control and Game Models of the Greenhouse Effect. XI, 225 pages. 1994.

Vol. 417: B. Ran, D. E. Boyce, Dynamic Urban Transportation Network Models. XV, 391 pages. 1994.

Vol. 418: P. Bogetoft, Non-Cooperative Planning Theory. XI, 309 pages. 1994.

Vol. 419: T. Maruyama, W. Takahashi (Eds.), Nonlinear and Convex Analysis in Economic Theory. VIII, 306 pages. 1995.

Vol. 420: M. Peeters, Time-To-Build. Interrelated Investment and Labour Demand Modelling. With Applications to Six OECD Countries. IX, 204 pages. 1995.

Vol. 421: C. Dang, Triangulations and Simplicial Methods. IX, 196 pages. 1995.

Vol. 422: D. S. Bridges, G. B. Mehta, Representations of Preference Orderings. X, 165 pages. 1995.

Vol. 423: K. Marti, P. Kall (Eds.), Stochastic Programming. Numerical Techniques and Engineering Applications. VIII, 351 pages. 1995.

Vol. 424: G. A. Heuer, U. Leopold-Wildburger, Silverman's Game. X, 283 pages. 1995.

Vol. 425: J. Kohlas, P.-A. Monney, A Mathematical Theory of Hints. XIII, 419 pages, 1995.

Vol. 426: B. Finkenstädt, Nonlinear Dynamics in Economics. IX, 156 pages. 1995.

Vol. 427: F. W. van Tongeren, Microsimulation Modelling of the Corporate Firm. XVII, 275 pages. 1995.

Vol. 428: A. A. Powell, Ch. W. Murphy, Inside a Modern Macroeconometric Model. XVIII, 424 pages. 1995.

Vol. 429: R. Durier, C. Michelot, Recent Developments in Optimization. VIII, 356 pages. 1995.

Vol. 430: J. R. Daduna, I. Branco, J. M. Pinto Paixão (Eds.), Computer-Aided Transit Scheduling. XIV, 374 pages. 1995.

Vol. 431: A. Aulin, Causal and Stochastic Elements in Business Cycles. XI, 116 pages. 1996.

Vol. 432: M. Tamiz (Ed.), Multi-Objective Programming and Goal Programming. VI, 359 pages. 1996.

Vol. 433: J. Menon, Exchange Rates and Prices. XIV, 313 pages. 1996.

Vol. 434: M. W. J. Blok, Dynamic Models of the Firm. VII, 193 pages. 1996.

Vol. 435: L. Chen, Interest Rate Dynamics, Derivatives Pricing, and Risk Management. XII, 149 pages. 1996.

Vol. 436: M. Klemisch-Ahlert, Bargaining in Economic and Ethical Environments. IX, 155 pages. 1996.

Vol. 437: C. Jordan, Batching and Scheduling. IX, 178 pages. 1996.

Vol. 438: A. Villar, General Equilibrium with Increasing Returns. XIII, 164 pages. 1996.

Vol. 439: M. Zenner, Learning to Become Rational. VII, 201 pages. 1996.

Vol. 440: W. Ryll, Litigation and Settlement in a Game with Incomplete Information. VIII, 174 pages. 1996.

Vol. 441: H. Dawid, Adaptive Learning by Genetic Algorithms. IX, 166 pages.1996.

Vol. 442: L. Corchón, Theories of Imperfectly Competitive Markets. XIII, 163 pages. 1996.

Vol. 443: G. Lang, On Overlapping Generations Models with Productive Capital. X, 98 pages. 1996.

Vol. 444: S. Jørgensen, G. Zaccour (Eds.), Dynamic Competitive Analysis in Marketing. X, 285 pages. 1996.

Vol. 445: A. H. Christer, S. Osaki, L. C. Thomas (Eds.), Stochastic Modelling in Innovative Manufactoring. X, 361 pages. 1997.

Vol. 446: G. Dhaene, Encompassing. X, 160 pages. 1997.

Vol. 447: A. Artale, Rings in Auctions. X, 172 pages. 1997.

Vol. 448: G. Fandel, T. Gal (Eds.), Multiple Criteria Decision Making. XII, 678 pages. 1997.

Vol. 449: F. Fang, M. Sanglier (Eds.), Complexity and Self-Organization in Social and Economic Systems. IX, 317 pages, 1997.

Vol. 450: P. M. Pardalos, D. W. Hearn, W. W. Hager, (Eds.), Network Optimization. VIII, 485 pages, 1997.

Vol. 451: M. Salge, Rational Bubbles. Theoretical Basis, Economic Relevance, and Empirical Evidence with a Special Emphasis on the German Stock Market.IX, 265 pages. 1997.

Vol. 452: P. Gritzmann, R. Horst, E. Sachs, R. Tichatschke (Eds.), Recent Advances in Optimization. VIII, 379 pages. 1997.

Vol. 453: A. S. Tangian, J. Gruber (Eds.), Constructing Scalar-Valued Objective Functions. VIII, 298 pages. 1997.

Vol. 454: H.-M. Krolzig, Markov-Switching Vector Autoregressions. XIV, 358 pages. 1997.

Vol. 455: R. Caballero, F. Ruiz, R. E. Steuer (Eds.), Advances in Multiple Objective and Goal Programming. VIII, 391 pages. 1997.

Vol. 456: R. Conte, R. Hegselmann, P. Terna (Eds.), Simulating Social Phenomena. VIII, 536 pages. 1997.

Vol. 457: C. Hsu, Volume and the Nonlinear Dynamics of Stock Returns. VIII, 133 pages. 1998.

Vol. 458: K. Marti, P. Kall (Eds.), Stochastic Programming Methods and Technical Applications. X, 437 pages. 1998.

Vol. 459: H. K. Ryu, D. J. Slottje, Measuring Trends in U.S. Income Inequality. XI, 195 pages. 1998.

Vol. 460: B. Fleischmann, J. A. E. E. van Nunen, M. G. Speranza, P. Stähly, Advances in Distribution Logistic. XI, 535 pages. 1998.

Vol. 461: U. Schmidt, Axiomatic Utility Theory under Risk. XV, 201 pages. 1998.

Vol. 462: L. von Auer, Dynamic Preferences, Choice Mechanisms, and Welfare. XII, 226 pages. 1998.

Vol. 463: G. Abraham-Frois (Ed.), Non-Linear Dynamics and Endogenous Cycles. VI, 204 pages. 1998.

Vol. 464: A. Aulin, The Impact of Science on Economic Growth and its Cycles. IX, 204 pages. 1998.

Vol. 465: T. J. Stewart, R. C. van den Honert (Eds.), Trends in Multicriteria Decision Making. X, 448 pages. 1998.

Vol. 466: A. Sadrieh, The Alternating Double Auction Market. VII, 350 pages. 1998.

Vol. 467: H. Hennig-Schmidt, Bargaining in a Video Experiment. Determinants of Boundedly Rational Behavior. XII, 221 pages. 1999.

Vol. 468: A. Ziegler, A Game Theory Analysis of Options. XIV, 145 pages. 1999.

Vol. 469: M. P. Vogel, Environmental Kuznets Curves. XIII, 197 pages. 1999.

Vol. 470: M. Ammann, Pricing Derivative Credit Risk. XII, 228 pages. 1999.

Vol. 471: N. H. M. Wilson (Ed.), Computer-Aided Transit Scheduling. XI, 444 pages. 1999.

Vol. 472: J.-R. Tyran, Money Illusion and Strategic Complementarity as Causes of Monetary Non-Neutrality. X, 228 pages. 1999.

Vol. 473: S. Helber, Performance Analysis of Flow Lines with Non-Linear Flow of Material. IX, 280 pages. 1999.

Vol. 474: U. Schwalbe, The Core of Economies with Asymmetric Information. IX, 141 pages. 1999.

Vol. 475: L. Kaas, Dynamic Macroeconomics with Imperfect Competition. XI, 155 pages. 1999.

Vol. 476: R. Demel, Fiscal Policy, Public Debt and the Term Structure of Interest Rates. X, 279 pages. 1999.

Vol. 477: M. Théra, R. Tichatschke (Eds.), Ill-posed Variational Problems and Regularization Techniques. VIII, 274 pages. 1999.

Vol. 478: S. Hartmann, Project Scheduling under Limited Resources. XII, 221 pages. 1999.

Vol. 479: L. v. Thadden, Money, Inflation, and Capital Formation. IX, 192 pages. 1999.

Vol. 480: M. Grazia Speranza, P. Stähly (Eds.), New Trends in Distribution Logistics. X, 336 pages. 1999.

Vol. 481: V. H. Nguyen, J. J. Strodiot, P. Tossings (Eds.). Optimation. IX, 498 pages. 2000.

Vol. 482: W. B. Zhang, A Theory of International Trade. XI, 192 pages. 2000.

Vol. 483: M. Königstein, Equity, Efficiency and Evolutionary Stability in Bargaining Games with Joint Production. XII, 197 pages. 2000.

Vol. 484: D. D. Gatti, M. Gallegati, A. Kirman, Interaction and Market Structure. VI, 298 pages. 2000.

Vol. 485: A. Garnaev, Search Games and Other Applications of Game Theory. VIII, 145 pages. 2000.

Vol. 486: M. Neugart, Nonlinear Labor Market Dynamics. X, 175 pages. 2000.

Vol. 487: Y. Y. Haimes, R. E. Steuer (Eds.), Research and Practice in Multiple Criteria Decision Making. XVII, 553 pages. 2000.

Vol. 488: B. Schmolck, Ommitted Variable Tests and Dynamic Specification. X, 144 pages. 2000.

Vol. 489: T. Steger, Transitional Dynamics and Economic Growth in Developing Countries. VIII, 151 pages. 2000.

Vol. 490: S. Minner, Strategic Safety Stocks in Supply Chains. XI, 214 pages. 2000.

Vol. 491: M. Ehrgott, Multicriteria Optimization. VIII, 242 pages. 2000.

Vol. 492: T. Phan Huy, Constraint Propagation in Flexible Manufacturing. IX, 258 pages. 2000.

Vol. 493: J. Zhu, Modular Pricing of Options. X, 170 pages. 2000.

Vol. 494: D. Franzen, Design of Master Agreements for OTC Derivatives. VIII, 175 pages. 2001.

Vol. 495: I Konnov, Combined Relaxation Methods for Variational Inequalities. XI, 181 pages. 2001.

Vol. 496: P. Weiß, Unemployment in Open Economies. XII, 226 pages. 2001.

Vol. 497: J. Inkmann, Conditional Moment Estimation of Nonlinear Equation Systems. VIII, 214 pages. 2001.

Vol. 498: M. Reutter, A Macroeconomic Model of West German Unemployment. X, 125 pages. 2001.

Vol. 499: A. Casajus, Focal Points in Framed Games. XI, 131 pages. 2001.

Vol. 500: F. Nardini, Technical Progress and Economic Growth. XVII, 191 pages. 2001.

Vol. 501: M. Fleischmann, Quantitative Models for Reverse Logistics. XI, 181 pages. 2001.

Vol. 502: N. Hadjisavvas, J. E. Martínez-Legaz, J.-P. Penot (Eds.), Generalized Convexity and Generalized Monotonicity. IX, 410 pages. 2001.

Vol. 503: A. Kirman, J.-B. Zimmermann (Eds.), Economics with Heterogenous Interacting Agents. VII, 343 pages. 2001.

Vol. 504: P.-Y. Moix (Ed.),The Measurement of Market Risk. XI, 272 pages. 2001.

Vol. 505: S. Voß, J. R. Daduna (Eds.), Computer-Aided Scheduling of Public Transport. XI, 466 pages. 2001.

Vol. 506: B. P. Kellerhals, Financial Pricing Models in Continuous Time and Kalman Filtering. XIV, 247 pages. 2001.

Vol. 507: M. Koksalan, S. Zionts, Multiple Criteria Decision Making in the New Millenium. XII, 481 pages. 2001.

Vol. 508: K. Neumann, C. Schwindt, J. Zimmermann, Project Scheduling with Time Windows and Scarce Resources. XI, 335 pages. 2002.

Vol. 509: D. Hornung, Investment, R&D, and Long-Run Growth. XVI, 194 pages. 2002.

Vol. 510: A. S. Tangian, Constructing and Applying Objective Functions. XII, 582 pages. 2002.

Vol. 511: M. Külpmann, Stock Market Overreaction and Fundamental Valuation. IX, 198 pages. 2002.

Vol. 512: W.-B. Zhang, An Economic Theory of Cities.XI, 220 pages. 2002.

Vol. 513: K. Marti, Stochastic Optimization Techniques. VIII, 364 pages. 2002.

Vol. 514: S. Wang, Y. Xia, Portfolio and Asset Pricing. XII, 200 pages. 2002.

Vol. 515: G. Heisig, Planning Stability in Material Requirements Planning System. XXX, 000 pages. 2002.

Lecture Notes in Economics and Mathematical Systems 514

Founding Editors:

M. Beckmann
H. P. Künzi

Managing Editors:

Prof. Dr. G. Fandel
Fachbereich Wirtschaftswissenschaften
Fernuniversität Hagen
Feithstr. 140/AVZ II, 58084 Hagen, Germany

Prof. Dr. W. Trockel
Institut für Mathematische Wirtschaftsforschung (IMW)
Universität Bielefeld
Universitätsstr. 25, 33615 Bielefeld, Germany

Co-Editors:

C. D. Aliprantis, Dan Kovenock

Editorial Board:

P. Bardsley, A. Basile, M.R. Baye, T. Cason, R. Deneckere, A. Drexl, G. Feichtinger, M. Florenzano, W. Güth, K. Inderfurth, M. Kaneko, P. Korhonen, W. Kürsten, M. Li Calzi, P. K. Monteiro, Ch. Noussair, G. Philips, U. Schittko, P. Schönfeld, R. Selten, G. Sorger, R. Steuer, F. Vega-Redondo, A. P. Villamil, M. Wooders

Springer
Berlin
Heidelberg
New York
Barcelona
Hong Kong
London
Mailand
Paris
Tokyo